Servicing ITSM

A Handbook of Service Descriptions
for IT Service Managers and a
Means for Building Them

Randy A. Steinberg

North America & international
toll-free: 1 888 232 4444 (USA & Canada)
fax: 812 355 4082

Other books by Randy A. Steinberg:

Implementing ITSM
From Silos To Services—Transforming The IT Organization
To An IT Service Management Valued Partner
Trafford Press ISBN: 978-1-4907-1958-0

Measuring ITSM
Measuring, Reporting, and Modeling the IT Service
Management Metrics that Matter Most to IT Senior Executives
Trafford Press ISBN: 978-1-4907-1945-0

Architecting ITSM
A Reference Of Configuration Items and Building Blocks For
A Comprehensive IT Service Management Infrastructure
Trafford Press ISBN: 978-1-4907-1957-3

Once Upon A Time . . .

. . . a large bank had one of the best IT departments in the world. They employed the smartest technologists who could build or fix anything. They put together the most reliable network. They had some of the best servers, error free applications, amazing storage networks and an internet support system that was the envy of all other companies.

One day, the king (actually the new CEO) of the bank began to question how IT actually served the bank. He simply asked: "What is it we are getting for all this money that we spend on our networks, our servers, our smart technologists and our error free applications?"

IT scratched their heads for a while. "Of course, the new CEO misunderstands us" they pondered. "He doesn't realize how complex our operation is, how hard our people work or how well IT does at keeping everything running." They decided to send their best and wisest technologist to talk to the CEO to explain this value in simple terms.

The new CEO was not satisfied with the IT response. "I need to know how IT contributes to our bottom line."

Frustration rose among both IT and the new CEO. To break the stalemate, the wise technologist suddenly remembered. "Of course", he said, "one of the things IT delivers is an online banking service that all the account holders in the land can use to check on their account balances, transfer money and pay their bills". "Of course!" said the CEO, "This is what brings customers to our bank instead of our competitors. How much do we spend on this service?" The wise technologist had no ready answer. "I'll go find out" he said.

And find out he did. He tallied the costs for all the people that supported the online banking service. He tallied all the hardware, the software, the storage, the networking and operational costs for running the service. He came back to the CEO and proudly announced what IT was actually spending $22.8M on the service per year.

"That's a lot of money", said the CEO. "How many account holders actually use that service?"

The wise technologist pondered for a while. He thought about the 4,000 account holders that subscribed to the service. "I guess this comes to $5,700 per year for each account holder."

The CEO was not happy. "Our people tell me that the average account holder holds deposits of $1,800. You mean that IT is spending $5,700 per account holder per year to maintain accounts with an average yearly balance of $1,800?"

"Oh well", thought the technologist, "at least we provide good availability. The network has few outages. So do the servers, the storage networks, the applications and the infrastructure. I see those reports every month."

Turns out the customer account holders didn't quite agree. While IT looked at the availability of each and every component in the infrastructure that provided the service, no one looked at the *combined effect* that all those little outages had on the customers. Suddenly, the wise technologist didn't look quite so wise.

Taking a service view suddenly exposed how poorly IT was really managing itself.

This was not really a fable. This actually happened to a major retail bank in Chicago.

Once exposed, the IT division of the bank focused on architecting their online banking service to reduce infrastructure costs. This resulted in freeing up significant amounts of working capital. It allowed the bank to considerably grow their online services at a lower per unit cost for each account holder. And this was only one of many services that IT delivered.

None of this would have happened if IT didn't step back to understand what services they actually deliver and how those services contribute to the bottom line of the bank.

IT typically cannot articulate what services they provide or how these contribute to the corporate bottom line.

Significant cost and quality issues become suddenly obvious once services are identified and a service view is taken.

. . . and that's why this book was written . . .

Dedication

This book is dedicated to those very hard working IT professionals, managers and executives who deserve to see their IT solutions deploy and operate day-to-day within acceptable levels of costs and risks to their company.

Contents

Chapter 4

Technical Management Services

Chapter 5

Operational Management Support Services

Chapter 8

Chapter 9

Chapter

1

Benefits of an IT Service Approach

Benefits of Managing By Services

IT is at the crossroads.

Companies can continue to manage themselves by technology silos as they have done for many years, or they can begin transitioning to an approach where IT manages itself by the services it actually delivers. This is because the business doesn't want the technologies—it wants the services.

If this management shift doesn't change, then continue to expect the same continuous rise in service disruptions and labor costs. Continue to expect increases in the amount of non-discretionary spend just to keep the IT wheels running.

Managing by silos is not wrong, just a practice that no longer works in today's world. The pace of technology and business change is moving too fast. The number of platforms and technologies is exploding. Virtualization, mobile devices, cloud computing, new software delivery technologies are examples of events forcing radical changes to the IT delivery model.

The services needed by the businesses today require that many disparate technologies must work together to achieve business goals. Yet, most IT organizations hide their heads in the sand by continuing to operate in technical silos—hoping that somehow, people will just do the right thing, work together and make it happen. No one is rewarded for doing so and no one is held accountable or responsible for the actual services being delivered.

The role of IT has primarily changed—from a primary focus on engineer-and-build to a primary focus on integrating services from many providers!

How can any IT organization succeed without an understanding of what it has to really provide? The business organization in any company wants services that provide value to the company. It does not care greatly about what technologies it is operating with. IT cannot begin to provide value that is recognized by the business until it recognizes, organizes and operates the services that the business needs.

Senior executives in a company understand little about IT technologies and how they are supposed to operate. They do understand that IT is expensive, prone to many disruptions and usually appears to run into the same kinds of problems repeatedly.

To Senior Management, IT represents a large non-discretionary fixed cost overhead of which about two thirds is needed "just to keep the wheels running". Many would like to reduce this overhead, or at least, shift the costs towards more value added activities such as new service solutions that may increase market share or profits.

With the current technology silo management approach, IT has effectively locked itself out of any meaningful discussions with company executives for becoming a partner with the business. In effect, managing by technical silos places IT in the position of being an overhead operation whose costs are required just to keep things running. New demands are placed on IT year after year many times without needed business support or funds simply ". . . because IT manages to get it done somehow . . ."

With a service management approach, suddenly many things start to change. The business, as well as IT itself, starts to look at things in a new light. Suddenly, new ways of operating more effectively are discovered. New priorities are seen for where to invest costs. Costs are actually reduced and the overall quality of services goes up. IT becomes a valued partner with the business in adapting to new competitive challenges and proactively building strategies—instead of just reacting to them.

For those not yet fully convinced about the power of the service approach, here are a few examples that have actually occurred with some real business organizations:

Large Global Media Services Company

This company was struggling with cost issues. They operated about 192 data centers in 76 countries around the world. In looking to cut IT costs, they initially tried getting each center director to identify areas for cost reductions. Many directors came up with small cuts. Some rebelled outright by saying they were too critical to business operations to reduce their costs any further. It was this latter complaint that led to looking at things differently.

Rather than ask each director to identify cost cuts, corporate management had them identify what services they actually delivered. A company auditor, outside of IT, didn't spend time auditing IT finances or trying to look at technologies. Instead, they asked: "What services are being delivered here?"

They then compiled a master list of services across all 192 centers and came up with some astounding comparisons. Many of the centers were delivering similar services that could be combined or delivered regionally. A key finding was that almost 40% of the centers mostly provided basic monitoring and maintenance services.

It didn't take a lot of expertise to quickly determine that significant savings could be had simply by centralizing some of these basic functions. What was the result? It allowed the company to close 70 data centers with no disruption to business operations.

Note that this approach was fully understandable to both IT as well as the business. It allowed the business to step in an overcome the protectionist behavior of the data center directors. It allowed the business to determine which services were important and which could be combined or operated differently. In the end, significant savings were obtained with almost no disruption to ongoing business operations.

National Banking Corporation

This company had expenditure issues. Major fixed non-discretionary costs were being sunk year after year into ongoing IT operations just to keep the bank running. The business was very concerned, not so much about the costs, but about a changing competitive landscape that required banking services to be delivered through different channels. IT was looking at significant cost increases to support them.

The bank put together a joint IT and business team to look at the problem. Again, the business asked: "What services are being delivered here?" They also asked two additional questions: "Which banking business functions do those services support?", and, "How much does the bank spend on each service?" The results again were quite interesting.

The bank discovered that most of its costs were going towards IT services that were supporting business channels that were non-strategic to the bank. "Why are we investing so much money year after year", they asked, "in areas that we really don't care about?" They created a quadrant showing each IT service and began dividing those services into strategic, non-strategic, high-value and low-value. This became a key tool for executives to identify where they wanted to shift IT costs to better compete for banking services.

Mid-sized Hospital in the U.S

This is an interesting example of how a service approach can motivate IT staff. For many years, this hospital operated nightly batch job runs to process billings and customer payments. Due to many architecture and operational issues, customer payment and billings were not catching up with hospital service costs. The hospital was starting to run millions of dollars in the red caused by the slowness in revenue collections.

To make matters worse, IT staffs at the hospital were unionized and many employees had been operating the systems there for many years. Changes in operations and systems were needed, but there was little IT incentive to change anything—until a service approach was accidentally introduced.

With a change in IT management, it was suddenly asked: "What services are being provided here?" in reference to the nightly batch runs. A business unit manager was forced to explain that each night, because of certain batch work, the hospital was able to collect revenue. "In fact", claimed the manager, "you can see exactly how much revenue was collected that night by looking in a particular log file at the end of the job run", a capability no one was previously aware of.

Surprisingly, this held great interest for IT staff that had operated these batch runs night after night for many years. They posted the dollars collected each night after the end of the batch runs. They quickly discovered that if the batch runs ran successfully, more revenue was collected than if they had problems. IT staff was amazed at how much money was involved. "You mean we collected over $5,000,000 just in one night?" they asked, "Wow! I didn't think we were so important!"

The impact of this knowledge was amazing. Suddenly, unionized IT operators were congratulating each other when they saw how much money was collected. "What can we do to make this number bigger?" they asked. One night, an application analyst put in a program change that was not fully tested or authorized. It caused a problem that made the number go lower. The disappointment registered throughout the IT staff. "We've got to make sure this never happens again", they cried.

Then the architects and programmers stepped in. "How can we really make this number go up?" they asked. Application, database and operational changes were made. The number kept going up night after night. To the operators and the IT staff, this was no longer just a set of applications, jobs and technologies that needed to be run each night—it became a service mission to collect as much revenue as possible, fully recognized by hospital management.

Within several months, the hospital was running over $2,000,000 in the black.

Global Financial Services Firm

This large firm had fixed annualized network capacity costs that were very expensive each year. Capacity planners at this institution simply examined network traffic trends carefully for each year and then doubled the bandwidth needed for the following year. When a new capacity planning manager came in, it was asked: "Why do you forecast this way". The answer: "Historically, we have no network capacity issues when we do this".

The new manager applied a service approach. "Why not examine the network traffic by the services that use it and then forecast based on those service needs?" The answer came back: "This is the way we have done it for many years. We have never had a capacity problem. It is too much effort to spend time looking at the use of the network and collecting individual forecasts that much more thoroughly".

Undaunted, the new manager started asking: "What services are we delivering here?" and, "How much are we spending on those services?"

The service costs were astounding—about $800 million U.S. dollars, a year just for line costs. It did not take long to figure out the following:

- Line costs were $800 million

- Capacity was doubled to "prevent issues"

- Doubling for capacity represented $400 million annually for line bandwidth of which much was unused

- A small 5% gain was worth $20 million annually

- For $20 million, the firm could hire a lot of capacity analysts to examine services more closely (which they actually did!)

National Pharmaceutical Distribution Company

This company looked at storage costs by identifying each service first, then mapping those services to the storage they consumed. Results of identifying unneeded services, low value services or which services weren't using their storage: about $25-30M in annual cost savings!

Large Parts Manufacturing and Distribution Company

This company also suffered from high operating costs. They also ran a highly problematic IT operation with numerous and frequent outages. "We've got way too complex an IT operation—we're not like anyone else", they said. That is, until a service view was taken.

It turned out that the company IT organization was really providing just 11 basic services that were needed by the business. So why were costs so high and operations so complex?

It had something to do about running 11 basic services in myriads of ways using all kinds of varying technology platforms. They were right about the complexity—wrong in that complexity was really called for in the first place.

Large National Mortgage Company

This company spent over $60,000,000 U.S. to establish their own service continuity strategy by building a new data center. In the end, they successfully proved that the mainframes running in each center could switch to the new center if a major disaster occurred.

What was little known, unless you took a service view, was that the mainframe could stay operational, but the national lending service would still be down because the mid-range servers and local area networks also needed to operate the service were not considered in the recovery plan.

This fact is still hidden from company executive management even today.

*　*　*

Many times, there are frequent complaints on behalf of IT for cost justifying an IT Service Management initiative. It could easily be argued that the answer is right there in front of everyone.

But it has to start with an understanding of what IT services are being delivered—because in the end—this is all the business really cares about.

IT Challenges With Services

One of the key challenges to IT is that it cannot effectively articulate what services it actually delivers. This may sound strange, but it is true. IT, for the most part, is so enmeshed with the technologies it needs to support that it frequently loses sight of the big picture.

Here is a list of responses given by real IT practitioners when asked "What services do you deliver?"

"We deliver CISCO and IBM . . ."

"We provide an object broker service used throughout many of the applications in our company . . ."

"We provide anything the business units ask for that has to do with IT (except certain things . . .)"

"We serve up web content to our IT subscribers . . ."

"I handle the mainframe . . ."

"We deliver database services . . ."

"We support the SAP application"

"We monitor and manage the servers . . ."

"We operate 20 call centers across the U.S. and Canada!"

"We run a data center for the company in Kentucky . . ."

"Don't ask me—it's probably someone else's job—I just take care of the network . . ."

The true purpose of this book is to cut through this misunderstanding on behalf of IT. Most organizations deliver a standard general set of IT services. This book identifies what those are. It is up to the reader as to whether to use those services as explicitly stated or whether to mix match or combine them into something that will work within their own organizations.

Other challenges IT will face in starting on a service approach:

- Confusion about what a service is

- Tendency to call almost everything a "service"

- Tendency to mix up services with the organizations that might deliver them (e.g. Unit A does X and Unit B does Y—therefore there must be two services—instead of maybe one service that provides the functions delivered by both X and Y)

- Reluctance to spend time on identifying services since there may be no reward for doing so

- Identify services only for the IT unit they work for which may be different than what the business really is looking for

- Confuse services with the processes that underpin them

- Confuse use of tools thinking that the tool is the service

- Confuse services with how they are delivered— typically create multiple services for each delivery channel instead of developing one service that is provided to all channels

- Delay any work on identifying services until
  ". . . we get to ITSM Service Level Management next
  year . . ."—a key mistake because services underpin
  every ITSM process

- IT operations may have little communications with
  applications development or poor communications
  with company business units.

- No one really has the big picture.

A secondary purpose of this book is to provide enough
information to hopefully overcome some of these challenges.

Book Chapters in Brief

Brief descriptions of the remaining book chapters are as follows:

Chapter 2—ITSM Service Basics

This chapter presents a high level overview of what an IT service is. It provides some guidance on identifying services and creating service descriptions. It also provides some advice for setting service targets and establishing metrics.

Chapter 3—Categorizing Services

This chapter covers a recommended set of service lines for your IT services. Service lines are used to help organize and categorize the many IT services that typically exist in most organizations.

Chapters 4 to 10—Service Description Chapters

These chapters list each IT service typically delivered by IT organizations. Each chapter represents one service line and all the services that belong to it. The services themselves have fairly detailed function and feature descriptions that can be used to start your own ITSM Service Portfolio and Service Catalog.

Chapter 11—Governing Services

This chapter presents a working governance process that can be used to manage and oversee services once they have been put into place. Included are some basic organizational roles for governance, typical activities and a four stage governance process that can be used.

Chapter 12—Service Implementation

This chapter presents a detailed service implementation plan that you can use to build your own services with. The plan is generic in that almost every service will need to go through similar steps. You can leverage this by adding detail specific to the services you plan to implement.

Chapter

2

ITSM Service Basics

What Is An IT Service?

A service is simply anything of value given to a customer.

As an example, if you are given a pair of shoes, you have not been given a service. If the shoes are shined, laces tied, heels repaired and placed on your feet for you, a service will have been rendered. Therefore, simply providing an IT technology, says a server for example, is not providing a service. If that server is maintained, repaired and managed, then a service has been given.

Services are associated with a value. Painting servers with new colors each month is providing a service, but will be seen as having little value to the business. Operating and maintaining those servers to support business operations most likely will be seen by the business as providing value. Value may change over time as business and competitive needs change. Therefore, it is important to maintain a portfolio of services and continually review their value against the costs of providing them.

Services should not be confused with organizations or IT units within the company. For example, a NOC (Network

Operations Center) is not a service. The NOC itself could deliver a service such as Monitoring, Incident Response or Event Management services.

A service can be delivered through one or more IT or business units. Delivery of IT services can happen internally within the company, through an IT shared services organization or through an external third party vendor.

A Customer is an entity that will consume or receive a service. Customers have specific needs and outcomes that they expect when receiving a service. If expectations are not met, low satisfaction levels will occur no matter how well IT delivers the service. A successful outcome is usually attained if customer expectations are not only met but exceeded.

A Buyer is an entity that pays for the service. It is important to recognize the difference between the buyers and customers. It is possible to have very high customer satisfaction levels but low buyer satisfaction levels if the costs for service delivery and support are too high. It is also possible to have low customer satisfaction levels and high buyer satisfaction levels if cost savings are achieved. Be on the lookout for these two types of situations as they create many problems for IT staff and management.

Services have Service Assets. These are simply the Configuration Items (CIs) that make up a service. Types of these include CIs such as hardware, software, people, supporting processes, delivery policies, reports, information, service agreements, and anything else needed to deliver services successfully.

Services should be documented with their assets showing Service Configurations. These are like a bill-of-materials listing that you might find for a product at a manufacturing plant. The Service Configuration provides a hierarchical explosion view of how a service is put together in terms of its assets.

Services are delivered through Delivery Channels. These simply state to the customer how they will receive the service. As an example, an Email Messaging Service may deliver electronic messages via delivery channels such as PCs, Laptops, mobile devices and tablets.

Services may also have Prerequisites. These are requirements that a customer must have before the service can be delivered. For example, in order to receive the Email Messaging Service customers must first have acquired their delivery channel device (PC, Laptop, or device), gotten an IP address and a management signed authorization form.

A Service Feature is an intended element of the service provided to the customer. A service will have one or more features that describe in lower level detail, what the customer will get with the service. Features for the Email Messaging Service, for example, might include storing and forwarding of electronic messages, distribution list management, 100MB of mailbox storage, calendaring and meeting scheduling capabilities.

Services are typically described in Service Portfolios and Service Catalogs. The Service Portfolio is inwardly focused to IT and business management who will be providing the services. It lists each service along with key attributes such as delivery costs, gained revenue, and strategic importance. The Service Catalog is outwardly focused to the customer. It contains key attributes such as service features, who to contact to get the service, who to contact for help and any service charges.

Optionally, services can be grouped into Service Lines. A Service Line is simply a logical grouping of similar or related services. It provides easier access when looking for specific services. This is not unlike a chapter in a book or a tab in a product catalog. While not required, you may find that grouping your services into categories or lines of service makes it much easier to navigate to what someone may be looking for.

Putting All the Service Elements Together

Services consist of many elements. Requests, changes, service level agreements, charges are all examples. The following illustration is suggested means for structuring all of these together:

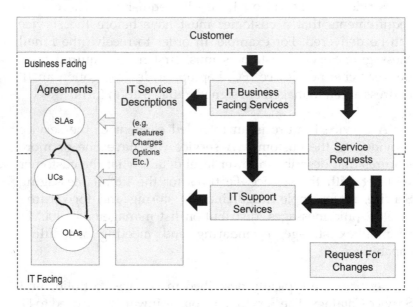

Figure 1: Putting Service Elements Together

At the very top we have the Customer. The Customer will order services to accomplish some business function. Those services that the customer actually sees and interacts with are referred to as IT Business Facing services. Examples might include an Accounts Payable Support Service, Order-To-Cash Support Service, Email, or Hosting Support.

The IT Business Facing services will typically require support services to fully operate. These services are referred to as IT Support Services. They are not seen directly by the customer, but are critical to successful operation of the IT Business Facing Services. Examples might include Database

Support, Server Administration, Monitoring, Backup, or Scheduling services.

Service Descriptions describe what (and not how) the customer will be provided with. A more detailed laundry list of what goes in service descriptions is provided later in this chapter. The descriptions include the features of the service, charges, options, delivery channels, availability, and delivery targets.

Service Descriptions are supported by agreements. These can be Service Level Agreements (SLAs), Operating Level Agreements (OLAs) and contracts with outside vendors called Underpinning Contracts (UCs). Of these, the Customer only sees the SLA. The SLA itself may be supported with OLAs and UCs.

Care should be taken that SLAs align with any OLAs and UCs that support it. An SLA that promises delivery of something in a 4 hour time period should be supported by the sum of all OLAs and UCs needed to make that delivery target.

Any IT Business Facing Service or IT Support Service includes a bundle of Service Requests. A Service Request can be viewed as a "transaction" against an IT service. It is the vehicle for how customers and users will interact with a service to obtain value. As an example, an Email service might include requests for adding new users, resizing mailboxes, creating distribution lists or removing users no longer with the company.

When designing and building a service, IT needs to get inside the head of those who will be using the service. What will they ask of the service? This means identifying all the possible types of requests that the service needs to include and ensuring the appropriate workflows and automation is in place to efficiently fulfill those requests.

Lastly we come to Requests for Changes (RFCs). Users and Customers interact with IT through service requests. The fulfillment actions taken for those requests may spin off RFCs as needed if those actions are creating changes to the IT infrastructure. RFCs are the vehicle by which IT support staff can make changes to the infrastructure and should remain invisible to the customer.

Service Ownership and Responsibilities

A key issue the IT Service Management addresses is to ensure accountability for the actual services being delivered to the customer versus the individual parts and technologies. A core problem in many IT organizations is accountability for technology silos (e.g. servers, databases, networks, applications) and no accountability for what all those items are being used for.

As an example, the IT department of a large retail chain might view IT success in terms of server uptime or network availability. The business may view IT success as number of inventory turns on a store shelf or IT cost per store sale. Without taking a holistic service view, IT is grossly missing the mark with the business and miscommunicating their value.

ITSM brings some new roles to the IT organization that serve to break the entrenched silos that might exist. Let's take a look at these:

Service Owner

Each service should have a Service Owner. This entity provides a single point of contact for the entire service and includes ownership for the service features, how it is delivered, costs and charges. The Service Owner also takes quality measurements of the service, reports on these and initiates improvement actions when deficiencies are found.

The Service Owner is also consulted on changes to the service that have been requested or proposed. In short, this role provides a single point of accountability for the overall delivery and quality of a service being delivered.

It is strongly suggested that Service Owners have a direct line to senior executive management.

Service Lead or Manager

Each service may also have a Service Lead or Manager. This role manages the delivery of the service on a day-to-day basis. It is responsible to ensure that the service is being delivered daily as expected and meeting service SLA targets.

Organizationally, this role can stay within the current technology silos if so desired but has at least a dotted line accountability to the Service Owner for each service supported.

Providers and Suppliers

Services will also have providers and suppliers. Providers represent the organization, either internal to the company or external, that directly provide the service to the customer. Suppliers provide support services and/or products to the provider organization that underpin the service being provided.

The relationship between the customer, the service provider and any suppliers is cemented using Service Level Agreements (SLAs), Operational Level Agreements (OLAs) and Underpinning Contracts (UCs). SLAs describe the agreement between the service provider and customer. The agreement between the supplier and the service provider is described with OLAs if the supplier is internal to the company or UCs if the supplier is an external 3rd party vendor.

Business Relationship Liaison

This role represents all IT services being provided to one or more business departments. It assists business customers with matching services to their needs and requirements. It reviews service delivery quality with key customer stakeholders. It may get involved with large development efforts representing the "voice of the customer" to the IT organization. Lastly, it serves to provide a readout of customer satisfaction back to the IT organization. It is strongly suggested that this role be independent of the IT organization or at least the IT technology silos that might exist.

Service Portfolios and Catalogs

There are two major artifacts used for communicating IT services. These are the IT Service Portfolio and IT Service Catalog. A simple way to distinguish between these:

- The IT Service Portfolio is inward facing to the IT organization and designed to manage IT investments, plans and strategies for the services being considered delivered or retired.

- The IT Service Catalog is outward facing to customers and designed to communicate the services available to them. The IT Service Catalog is generally a subset of the IT Service Portfolio.

The relationship between these is as shown below:

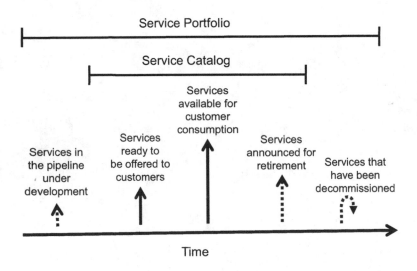

Figure 2: Service Portfolio To Service Catalog Scope

The IT Service Catalog will be used to describe your IT services to the business community. This should be put into business terms as much as possible. The Catalog should only have information about services that the business really needs to know. The Catalog itself can be delivered as a published document or website.

Keep it simple. Many companies that work with ITSM tend to make this a much bigger effort than needed. Basics such as service name, high level description, features, contacts and key delivery targets (in business terms) are all that is needed. If the catalog becomes too complex, technical or hard to read, the business won't use it.

A word about "Actionable" Service Catalogs. This term is generally used to describe service catalogs that also act as portals for users to access the requests that go with their services. Rather than merely provide a static link to a service description or document, these kinds of catalogs provide direct links to make service requests.

The IT Service Portfolio will be used by IT executive management as a tool for summarizing the services they are investing in. This portfolio should contain all the items found in the IT Service Catalog plus additional information about the services used to guide investment and improvement priorities and activities.

Identifying IT Services

Most IT organizations do not have the luxury of shutting down and retooling while they figure out which services they want to provide. Typically, they have already been delivering things for many years already, just without any service focus or discipline. Therefore, challenges will exist in taking what is being delivered today and repackaging that into service bundles and offerings.

Why bother? After all, if IT is delivering on things today, what's the point of recasting everything into services? The answer is that the current way of delivering is not sustainable for the future. Technology and the rapid pace of business change are moving so fast that those that cannot recognize the service end state of what they are delivering are at increasing levels of risk to properly integrate and deliver effectively to the business. As an example, if you are providing banking services to customers over a mobile device through a device app, over the internet, hosted by a 3rd party provider with data services running locally on a back-end virtualized server; and there is no overall accountability for that entire service; you are in deep trouble.

Therefore, IT must start by turning what is being delivered today into services. Do not fear. Almost every time this is met with support by the business. It is the IT support organizations that will present the most challenges. The attempt of this chapter is to help those in IT determine, from what they already have, what services they are really providing in a manner that is consumable and acceptable by business people.

An IT business facing service is a service that is based on the use of Information Technology to support a business service, business function or business process. Start by looking around your company. Which business activities appear to require the use of IT resources such as PCs, applications, networks or reports?

Examples might include:

- Email services that support communication throughout the company

- Applications that support corporate functions such as Payroll, HR, Finance and Accounting

- Applications that support payment processing, billings or collections for a service or product that the company delivers

- Hosting of servers or websites that communicate or sell goods and services to outside customers

A key purpose of this book is help you cut down on the research for what your services might be. Presented in later chapters, are descriptions of services commonly provided by most IT organizations. You can use these as a starting point, build on them or combine them based on the needs of your own company.

To begin your efforts in identifying what your business facing services might be, go after items such as:

- Listings of application inventories

- IT Service Continuity Plans to see what has been identified as vital business functions

- Company organization charts to see how business functions are being supported across the company

- Business process projects or other similar initiatives that have analysis already done for business processes and services

- Company annual reports, websites or other sources that advertise what the company delivers

Pull a small team together to sort through these with a series of working sessions. The team should be composed of representatives from IT operations and development units. IT service liaisons are also quite helpful as well as some representation from the business itself.

Application inventories are a key source for driving out the IT business facing services being provided. During the working sessions, walk through the inventories line by line. For Application A, for example, what business process, service or function does it support? Develop a service name for it. For example, if Application A supports the company Sales functions, then you might call the service something like Sales Support.

Now look at the next application. Would it also fall under the Sales Support service? The criteria for this might be that the application performs similar functions or is used by the same business unit. If not, create another service that it should fall under. Repeat this step for each application you come across in your inventory. This is illustrated by the following process flow:

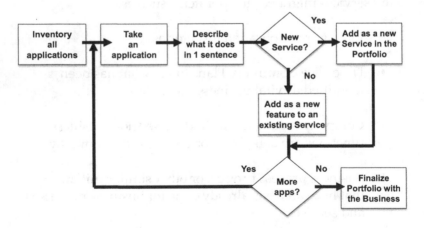

Figure 3: Process For Deriving Services From Applications

IT Service Continuity Plans are also a good source. These will indicate a more enterprise perspective of what the company sees as their services. It also points out which services are considered critical to company business operations.

Company organization charts are also helpful in that they provide clues as to what the business feels their business services really are. These would help point out business service areas such as Sales, Marketing, Finance or Manufacturing. Identified areas could then be translated into IT Business Support services such as Sales Support, Marketing Support, Finance Support and Manufacturing Support services.

Once your analysis efforts have completed, look at the number of services that were identified. Ideally, this number should be less than 80. More than that, you may be working at too fine a level of detail. Keep in mind that the more services you define, the more overhead and administration you will create in defining, managing, operating and reporting on services in your infrastructure.

Some companies like to brag how they operate with a list that numbers thousands of services. Observations have shown that these companies typically have a long list of services that no one pays much attention to. Try to keep your list as small as reasonably possible. A list of about 40-70 services for a mid-size to large business should be generally reasonable.

How do you pare down a large number of services? Take a second pass through your list with your team and look for opportunities to combine services. For example, you may have a list that shows support services for General Ledger, Accounts Receivable, and Accounts Payable. Consider combining all those services into a single Corporate Accounting Support service.

Having gone through this exercise, you will have completed a key step in starting down the ITSM IT Service Management path. You will identified your IT business facing services and features. You will have also identified some key application configuration items and associated them with the services you defined.

What about the IT support services? For these, many will already be described in this book. Feel free to mix and match them as needed to meet your own individual IT support organization requirements.

This is a great starting step towards beginning to configure your services and developing service models. The results also provide some immediate value in assisting impact analysis efforts for changes, incidents and new services.

Service Descriptions

Services should be described using a standard service description template common across all the services you have defined. The template consists of service attributes used to describe a service. All the template attributes should be in the IT Service Portfolio. A subset of them should be used for the IT Service Catalog.

An example list of possible service attributes that could be used for your template is described in the table below:

Table 1: Service Description Attributes

Service Attribute	Attribute Description	In IT Service Catalog?
Name	The name of a service.	Yes
Description	This is a one or two sentence high level description of the service stating what it provides to the business.	Yes
Category	The name of the Line of Service that this service will fall under.	Yes
Features	List key functions or features that will be delivered with the service.	Yes
Status	Whether the service is Proposed, Under Development, Operational, Retired or any other category you may wish to use.	Yes
Owner	Name or email contact of person with single point of accountability for the service.	Yes

Service Attribute	Attribute Description	In IT Service Catalog?
Availability	Describe when the service is available. E.g. 24 hours 7 days per week including holidays.	Yes
Support	Enter first level contact information for getting help and/or reporting problems with the service. This can be a Service Desk number, name, helpful web site, etc.	Yes
Initiated	Enter source for how the service can be obtained. E.g. website URL, contact number, Email address, etc.	Yes
Charges	List any charges and charging assumptions associated with receiving the service that would be paid by the customer.	Yes
Delivery Channels	Describe the means for how the service is delivered to the recipient. E.g. Desktop, laptops, Blackberry, hardcopy report, website posting, etc.	Yes
Pre-requisites	List any requirements that recipients must have in order to get or use the service such as owning a standard desktop, IP Address, approval from some source, etc.	Yes

Service Attribute	Attribute Description	In IT Service Catalog?
Customer	Indicate what business departments, cost centers, units or external customers are eligible to receive the service.	Yes
Technical Scope	Describe the platforms, applications, software or hardware that is covered within the scope of the service being provided.	Maybe
Suppliers	List the sources for those parties, internal or external to the company that assist in delivery or supply the service in some manner such as an external vendor, outsourcer or internal department or business unit.	No
Model	Put the link or reference to a CMDB, website or document that describes the underlying configuration for the service.	No
Class	Identify the classes of services available such as Gold, Silver, Bronze, VIP, etc. These may refer to delivery quality options for the service.	Yes
Artifacts	List any artifacts that a customer may receive as a result of getting the service such as a report, data repository, repaired item, etc.	Yes

Service Attribute	Attribute Description	In IT Service Catalog?
Key Performance Indicators	Identify quality measurements and assumptions that indicate how well the service is being delivered.	No
Key Service Targets	Identify high level service targets or objectives that set expectations about how the service will be delivered with the customer. These should be in business terms.	Yes
Key Benefits	List any key benefits or outcomes to the business or recipients because of receiving the service.	Maybe
Policies	Put the link or reference to a CMDB, website or document that describes the underlying policies and controls used for the service.	Maybe
Delivery Locations	List physical location addresses that would receive the service.	Maybe
Costs	Identify and list all infrastructure costs for operating, delivering and supporting the service.	No
Optional Features/ Charges	List any optional features and charges that may come with the service.	Yes
Service Level Agreements	Put the link or reference to a CMDB, website or document that describes the SLA associated with the service.	Yes

Service Attribute	Attribute Description	In IT Service Catalog?
Supplier Agreements	Put the link or reference to a CMDB, website or document that describes any Operational Level Agreements or Underpinning Contracts associated with the service.	No
Escalation Policy	Put the link or reference to a CMDB, website or document that describes the escalation policy to be used with the service.	Maybe
Constraints	List any constraints associated with the service such as transaction volume limits, limits on number of customers served or geographical constraints.	Maybe
Regulatory	List any external industry or government regulatory policies or controls associated with the service.	Maybe
Service Asset CI Types	List any CI types associated with delivering the service such as management, organization, processors, knowledge, people, etc.	No
Priority	Indicate the business priority level that the service may reside in. For example, the service may support a vital business function.	No

Service Attribute	Attribute Description	In IT Service Catalog?
Service Activation Period	Indicate the period of time that can be expected to activate the service for customers.	Maybe
Service Decommission Period	Indicate the period of time that can be expected to deactivate the service for customers.	Maybe

Not all of these attributes need to be used, but you should feel free to utilize those that best meet the needs of your efforts.

Service Targets and Metrics

Description of appropriate service levels, objectives or targets is one of those IT tasks that have been performed extremely poorly for many years and by many organizations. The core problem is this: IT confuses operational targets needed to deliver good services with service targets simply needed to meet business requirements.

Customers have surprisingly few and basic service targets they care about. They do not wish to see complex service reports showing things like server uptime, line utilizations, network traffic or transaction counts. Are these items important? They certainly are, but not to the business. IT typically does a poor job at separating out those targets needed to be seen by the business from those targets needed to perform or manage IT tasks day-to-day.

If you took your car into a mechanic's shop to be repaired because it wouldn't start that morning, how interested would you be in getting a multiple page report showing tire tread wear, valve and cylinder clearances, battery power draw voltage or engine oil pressure readings? These things are of interest to the mechanic providing the repair service, but certainly not of interest to you. You just want the car to start.

Here is a basic general short list of what customers typically want when it comes to IT services:

Customers prefer to get their own work done without stumbling over technology issues

The complexities of running an IT infrastructure make it easy for IT people to communicate to customers in technical terms and issues. In reality, most customers would prefer to avoid these and have little interest in them.

When things will fail, they expect quick recovery.

Customers actually understand that IT technologies are not always perfect and 100% available. Their expectations are that sometimes things will fail. When they do, they expect to see service restored as quickly as possible and need to feel confident that IT is responding with the appropriate effort and priority

They want to decide and don't-like hearing "no" from IT.

This does not mean that IT needs to give in to every demand. What it does mean is that IT needs to act like an advisor and partner—not as a barrier. Many times, there could be solid business reasons for asking for new or changed services.

When faced with a customer demand that appears to be a delivery challenge, inform the customer on how much their idea will cost. Let them know what the consequences of what doing it their way will be. Tell them what they may have to give up in order to have it their way. Then, let them decide for themselves whether that idea should be pursued or axed.

Better yet, implement a process to catch these kinds of requests and appropriately review and approve them. One such approach is in this book in the Governing Services chapter.

They dictate the service requirements, not IT.

This golden rule is very frequently ignored by most IT organizations. At the end of the day however, delivering to service requirements that the business neither cares for nor needs only diminishes the respect and capabilities in the eyes of the business. Nothing is gained by operating with targets that the business hasn't bought into and IT is just burying their head in the sand if they think otherwise.

A key mistake most IT organizations make is setting service levels and targets based on what they think they can deliver instead of asking what is needed directly from the customer. The fear is that they would somehow be committed to whatever the customer asks for.

Nothing is to be gained by having IT set its own targets. In fact, why would any respectable IT manager want to step out on the pier by themselves and set critical targets that may not be what the business needs?

Customer targets should be known by asking the customer directly and immediately. It is a waste of time and effort on everyone's part to "guess" what the correct service requirements are.

If it appears that targets cannot be made, then negotiations should occur to find a middle ground. If it is important enough, the business may even put up extra funds to get it achieved.

So what kind of service targets should be communicated to the business when it comes to IT services? Here is a short list of considerations:

Availability Service Targets

Nothing gets business executives, users and customers more upset than seeing service availability targets that are expressed in terms like "provide service availability of 99.8%". Expressing availability like this only serves to demonstrate to the business that IT is not in their camp when it comes to delivering the services they need. "Why is this not 99.9%" they might ask, or, "Why is it this number?" This kind of service target expression serves little purpose with the business.

There is nothing wrong with a service availability target expressed as the follows:

This service will be available from 6:00am to 10:00pm Monday through Friday except for holidays. If more than 4 service disruptions occur in any given month lasting more than 20 minutes in duration, a service investigation will be initiated to correct the deficiency.

Customers understand that IT technologies are not perfect. They just want to know that IT is on their side. They want to feel comfortable that IT will quickly and proactively address service deficiencies to the greatest extent possible. Expressing an availability target in this manner tells the business indirectly: "We know you need 100% availability and are committed to work our hardest to ensure you get it. If something goes wrong, we won't be waiting for a phone call—we will already be working diligently to correct the problem".

Service Penalties

Never put these into a service agreement unless absolutely pressed against the wall by a customer. Almost all situations where customers insist on including service penalties are those where there is little trust between the provider and the customer. No customer ever signs up for IT services hoping to get monies from penalties. They want the service.

One exception to this may be where a customer receives a true financial loss because of a disruption. This then becomes a negotiation. Even with this, IT should react like a business risk insurer. What additionally should be charged to cover the service risks?

Recovery Times

This is included for those customers that cannot sustain service outages beyond certain periods. Remember that the customer wants the service back. Restoring systems or equipment is meaningless to them if the service still is not useable.

Recovery times should not only refer to how quickly the service is restored if disrupted by an incident. Targets may also need to be set for recovery in the event of a major business disruption as well.

Avoid setting targets for recovery events that are meaningless if the event occurs but the service is still down. Typical examples include meaningless items such as:

- "Dispatch technician to arrive onsite within n hours . . ."

- "Replace failed component in x time . . ." (Where the service is still down even though the component has been successfully installed).

- "Return phone call in x minutes . . ."

Delivery Times

These include items such as delivery of reports, equipment, parts or other goods within specific timeframes. When needed, make sure there are agreed criteria for how delivery times will be measured, agreed and verified.

Response Times

These include items that might cover expected periods for responsiveness from the service. Examples of this might be items such as how quickly a repair technician shows up on site, how quickly the service desk answers a call, how quickly a transaction is processed or how quickly a new screen or web page might appear after the customer has hit their enter key.

Response times should not be set arbitrarily. First, agree that the requested time is critical to the needs of the customer. Then agree on the criteria by which the response time could be verified, measured and alerted to IT if missed.

Validating that response time service targets have been met can be done in a variety of ways. One approach is the hard core monitoring and measurement of everything in the path of the item being measured. Another approach is to agree on some relative indicator. For example:

"As long as updates to database ABC occur under 2 seconds, we will agree that response time objectives are being met . . ."

Yet another approach might be to conduct random audits and agree that objectives have been met if nothing fails the audit criteria. Still another approach is to create and operate simulated transactions that traverse the infrastructure. If these kick off or return within a specified timeframe, then it is agreed response time objectives are being met.

One last approach, although a bit reactive, might be to simply register and count complaint calls to the service desk. If they exceed a certain threshold, than a service investigation would be initiated.

Accuracy

Some services may require service targets around accuracy of data or information that is being provided. An example of this might be a Commodity Trading Support service that requires commodity prices to be accurate. Inaccurate prices might result in major financial loss if trades are done on the wrong information.

For accuracy targets, make sure that precise criteria for what represents accurate information is well defined. This may also include a tolerance range for accuracy as well as possible counts of inaccurate items.

IT may be surprised at how few type of targets exist that customers really care about. The details of infrastructure measurements such as server uptime, network latency, and equipment utilizations should be considered as operational information necessary to manage services so that they meet those customer service targets. They should not be part of a contracted Service Level Agreement with the customer.

For a serious look at service metrics, you may want to reference the **Measuring ITSM** book (ISBN: 978-1-4907-1945-0). This provides a comprehensive set of service metrics as well as access to a simple dashboard for reporting on them.

Chapter

3

Categorizing Services

Using Service Lines

It is highly recommended that you develop a small set of Service Lines to categorize the IT services that are being provided. A Service Line is merely a grouping of related services. Without this, you may have a long list of services that will be hard to wade through or communicate. Think of this as something like tabs or high level indexes that you might find in a product or retail catalog.

There is no set standard for what your Service Lines will be. In many cases, the business strategy, culture and operating style may influence what the Service Lines apply to your organization the most. The following illustrates one example which is used throughout this book:

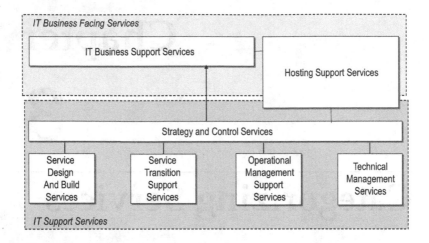

Figure 4: Example IT Service Lines

In the above model, the service lines have been organized into IT business facing services and IT support services. The business facing services are the ones that customers and users directly interact with. The IT support services are more in the background and seen mostly by IT operations, support and development units.

The Hosting Services line is used to handle bundles of internally or externally managed IT services usually delivered via cloud-based infrastructures. This can include items delivered "as a service" such as Software-as-a-Service (SaaS) or Platform-as-a-Service (PaaS) kinds of solutions. These can be provided directly to business customers or IT support staff.

It is the author's contention that the IT support services described in this book are common across most IT organizations regardless of industry. The IT business facing services may be unique to your company and industry. In reality, you may have multiple service lines that to categorize all your business facing services. As an example, it would not be unusual to have administrative, accounting, sales, marketing, and distribution service lines that contain IT business facing services.

The Service Lines examples used throughout the book as well as the services included within each line have been loosely based on an ITSM Service Lifecycle. Some liberties were taken with moving some of the services into the Strategy and Control Services service line. For example, this book views the management of IT changes more of a control service rather than a transition service.

Note that services are not processes. The relationship of ITSM processes to the services described in this book is that the processes are used to support those services. For example, the Database Management service in the Resource Management Services service line will most likely be supported by many ITSM processes such as Incident, Problem, Change, Asset and Configuration Management, etc.

You may think that some of the IT support services shown here look like common IT processes. A key concept to consider is this: is Incident Management (for example), a service, process or function? The answer is that it could be all three depending on your perspective. It is quite possible to have an IT support service called Incident Control, supported by the Incident Management process, and staffed by an Incident Management department.

The following sections describe each service line used in the book in more detail.

Technical Management Services

This Service Line includes services that implement, operate and maintain types of physical IT service assets that underpin IT and business services. Services in this line are grouped by technology platforms. A recommended set of services for this line are as follows:

- Server Management
- Database Management
- Application Management
- Network Management
- Storage Management
- Print Management
- Fax Management
- Physical Facilities Management
- Telephony Management
- Personal Computing Device Management
- Mobile Device Management
- Specialized Device Management
- Virtualization Management
- IT Supplies Management
- Middleware Transaction/Message Management

These kinds of services are probably the most familiar to IT technicians and support staff. They are described in much more detail in later in this book.

Operational Management Support Services

This Service Line includes services that manage and maintain IT operational workflows that cut across all the IT service assets in the physical infrastructure. Services in this line are grouped by those operational workflows that are necessary, but not directly supportive of business functions. A recommended set of services for this line are as follows:

- Service Desk
- Service Monitoring
- Incident Response
- Problem Control
- Request Fulfillment
- Backup/Restore Management
- Job Schedule Management
- Dispatch and Break-Fix Support
- Clock Management
- Service Startup/Shutdown Management
- File Transfer and Control Management
- Archive Management
- Data Entry Support
- Report Packaging and Distribution Support

These kinds of services are usually more familiar to IT operators and operational support staff. They provide service support needed to keep services running day-to-day. They are described in much more detail later in this book.

Service Transition Support Services

This Service Line includes services that provide support for transitioning new or changed service solutions to production operations. These services take a service solution or a service change and move it from a development state into an operational state. A recommended set of services for this line are as follows:

- Release Planning and Packaging
- Service Deployment and Decommission
- Site Preparation Support
- Service Validation and Testing Support
- Training Support
- Organizational Change Support
- Knowledge Management

These kinds of services are somewhat familiar to IT developers and operational support staff. They land somewhere in the middle between development and operations so service ownership and responsibility may be new to many IT organizations who typically fragment these services among many groups. They are described in much more detail later in this book.

Solution Design and Build Services

This Service Line includes services that plan, build and construct new services or changes to existing services. They take requirements for new services or service changes and translate these into service solutions. A recommended set of services for this line are as follows:

- Operational Planning and Consulting
- Solution Planning and Development
- Development Support Operations
- Capacity Management
- Availability Management
- Service Continuity Management
- Website Support

These kinds of services are generally familiar to IT developers and technical support staff. They also land somewhere in the middle between development and operations so service ownership and responsibility will also be somewhat new to many IT organizations. They are described in much more detail later in this book.

Strategy and Control Services

This Service Line includes services that govern and manage service delivery quality. It includes services that provide strategies for how services will be delivered. It also provides controls around their delivery to ensure that services are delivered safely and in compliance with industry and government regulations. A recommended set of services for this line are as follows:

- IT Service Strategy Support
- Architecture Management and Research
- IT Financial Management
- IT Project Management
- Change Control
- Configuration and Asset Management
- Lease and License Management
- Access and Security Management
- Service Audit and Reporting
- IT Workforce Management
- Procurement Support
- Process Management
- Supplier Relationship Management

These kinds of services are also generally familiar to IT developers and operational support staff. They are described in much more detail later in this book.

Hosting and Cloud Support Services

This Service Line includes bundles of IT support services that host IT functions. Services in this line can either serve customers directly or be used internally by IT support staff. Generally these kinds of services will use internal or external cloud-based delivery infrastructures. These might look like time sharing or outsourcing from a customer perspective. Examples of hosting support services could include:

- Basic Support Service
- Infrastructure As A Service (IaaS)
- Platform As A Service (PaaS)
- Application As A Service (AaaS)
- Software as a Service (SaaS)
- Network as a Service (NaaS)
- Security as a Service
- Storage as a Service
- Equipment as a Service
- Secure Controlled Infrastructure Facility (SCIF)

These kinds of services are also generally familiar to IT developers and operational support staff. They are described in much more detail later in this book.

IT Business Support Services

This Service Line includes IT business facing services that directly support company business functions, business partners and customers. They represent IT services that customers and users directly interact with.

Business facing services can vary greatly from company to company and industry to industry. Some services are supporting IT shared capabilities that business organizations use. Others may be specific to support of business functions and outcomes specific to the organizations that run the business of the company. Examples of IT shared services that businesses, no matter which industry they serve, might directly use could include:

- Desktop Support
- Data Warehousing and Business Intelligence
- Internet Telephony Service
- Email and Messaging
- Service Introduction
- Other IT Business Facing Services

The services in this line will vary greatly from company to company. For this reason, a chapter is included that describes a large sampling of typical services by company industry.

Chapter

4

Technical Management Services

This category includes provision of skills and services that implement, operate and maintain physical service hardware and systems software assets that underpin IT and business services.

- Server Management
- Database Management
- Application Management
- Network Management
- Storage Management
- Print Management
- Fax Management
- Physical Facilities Management
- Telephony Management
- Personal Computing Device Management
- Mobile Device Management
- Specialized Device Management
- Virtualization Management
- IT Supplies Management
- Middleware Transaction/Message Management

Server Management

Description:

Provides a service to implement operate and maintain physical and virtual server hardware and systems software assets that underpin IT and business services.

Service Scope:

This service manages server hardware, server operating systems and related systems software, customizing server hardware for network connectivity, loading and unloading application components to support services.

Service Functions and Features:

- Plan, install, configure and test server hardware and operating systems software configurations

- Configure servers for network access and perform connectivity testing to ensure servers are recognized by network

- Identify physical facility requirements needed to operate servers (i.e. floor space, equipment clearance, electrical, cooling, cabling, weight load)

- Decommission server hardware and related software upon request

- Provide server hardware, software and networking requirements to support procurement activities

- Install or remove service application and scripting components on request

- Maintain server hardware and software assets in compliance with supporting 3rd party vendor and lease requirements

- Maintain information about installed server hardware, software and networking configuration items and ensure accuracy and availability to others

- Label servers with asset tags and track server locations, serial numbers and owners

- Provide consulting services and support for release package testing, installation, deployment and operation

- Provide troubleshooting and technical support services for server hardware, software and networking components

- Implement capacity planning and tuning actions for server assets

- Coordinate and schedule server repair services with 3rd party vendors and validate that expected repairs and software patches achieved expected benefits

- Maintain Server Known Error and bug lists

- Provide consulting and support services to identify server operational, monitoring and reporting requirements

- Plan, architect and administer server virtualization solutions for server and operating system assets

Service Initiation:

- Approved Work Requests

- Escalated Incidents or Problems from the Service Desk

Service Delivery Channels:

- Satisfied Work Requests

- Consulting and Support

Database Management

Description:

Provides a service to implement operate and maintain database assets that support IT and business services.

Service Scope:

This service designs, builds, manages and maintains databases and large data repositories and stores used to support services.

Service Functions and Features:

- Plan, install, configure and test database configurations

- Identify server, network and capacity requirements needed to operate databases

- Decommission databases upon request

- Provide database requirements to support procurement activities

- Integrate databases with service application components on request

- Maintain database software assets in compliance with supporting 3rd party vendor requirements

- Maintain information about installed database software and database configuration items and ensure accuracy and availability to others

- Provide consulting services and support for release package testing, installation, deployment and operation

- Provide troubleshooting and technical support services for databases

- Coordinate, schedule, implement and test database tuning, reorganization, backup and restoration activities

- Provide consulting and support services to identify database operational, monitoring and reporting requirements

- Define physical and logical models for implementing and operating databases

- Identify database capacity sizing needs and requirements and manage database performance and tuning

- Define, implement and test database security schemas per requested requirements

- Maintain database Known Error and bug lists

- Plan, design, build, implement, test and maintain database checkpoint restart mechanisms

Service Initiation:

- Approved Work Requests

- Escalated Incidents or Problems from the Service Desk

Service Delivery Channels:

- Satisfied Work Requests

- Consulting and Support

Application Management

Description:

Provides a service to maintain and support application assets that underpin business services.

Service Scope:

This service manages and maintains application assets that have been created by internal development teams or packaged vendor products that are running in production.

Service Functions and Features:

- Plan, install, configure and test application minor upgrades, patches and fixes

- Identify server, network and capacity requirements needed to operate applications

- Decommission applications upon request

- Maintain 3rd party applications in compliance with supporting vendor requirements

- Maintain information about application configuration items and ensure accuracy and availability to others

- Provide consulting services and support for release package testing, installation, deployment and operation

- Provide troubleshooting and technical support services for applications

- Provide consulting and support services to identify application operational, monitoring and reporting requirements

- Coordinate, schedule, implement and test application tuning activities

- Identify requirements to support application sizing needs and requirements

- Define, implement and test application security schemas per requested requirements

- Maintain application Known Error and bug lists

- Plan, architect and administer service application virtualization solutions for application assets

Service Initiation:

- Approved Work Requests

- Escalated Incidents or Problems from the Service Desk

Service Delivery Channels:

- Satisfied Work Requests

- Consulting and Support

Network Management

Description:

Provides a service to maintain and support the networking and telecommunications infrastructure used to underpin IT and business services.

Service Scope:

This service manages and maintains the networking and telecommunications infrastructure that provides digital transport and connectivity across that infrastructure. This includes components such as routers, hubs, switches, lines, network appliances, as well as the overall WAN, local area networks, metropolitan networks and transport mechanisms like DSL and Broadband.

Service Functions and Features:

- Plan, install, configure and test networking infrastructure components and connectivity

- Identify network load, impact and capacity requirements needed to support services

- Identify physical facility requirements needed to operate networking components (i.e. floor space, equipment clearance, electrical, cooling, cabling, weight load)

- Provide networking requirements to support network procurement activities

- Decommission networking infrastructure components upon request

- Maintain networking components in compliance with supporting vendor requirements

- Maintain information about network topology and configuration items and ensure accuracy and availability to others

- Provide consulting services and support for release package testing, installation, deployment and operation

- Provide troubleshooting and technical support services for the networking infrastructure, IP addresses and naming services

- Provide consulting and support services to identify network operational, monitoring and reporting requirements

- Coordinate, schedule, implement and test network tuning activities

- Define, implement and test network security schemas per requested requirements

- Label network components with asset tags and track component locations, serial numbers, IP/MAC addresses and owners

- Coordinate and schedule server repair services with 3rd party vendors and validate that expected repairs and network patches achieved expected benefits

- Manage and maintain definitive hardware stores for networking spare parts and equipment

Service Initiation:

- Approved Work Requests

- Escalated Incidents or Problems from the Service Desk

Service Delivery Channels:

- Satisfied Work Requests

- Consulting and Support

Storage Management

Description:

Provides a service to maintain and support infrastructure digital storage assets used to underpin IT and business services.

Service Scope:

This service manages and maintains all forms of storage used across the enterprise. This includes storage area networks (SAN), attached storage, storage channels, tape devices, CD-ROM, microfiche, image devices, electronic storage, fiber channels, cache devices, hierarchical storage, virtual pools, and storage arrays.

Service Functions and Features:

- Plan, install, configure and test storage infrastructure components and connectivity

- Identify storage capacity requirements needed to support services

- Identify physical facility requirements needed to operate storage components (i.e. floor space, equipment clearance, electrical, cooling, cabling, weight load)

- Maintain and manage physical facilities for storage media such as tape libraries

- Provide storage requirements to support procurement activities

- Decommission storage infrastructure components upon request

- Maintain storage components in compliance with supporting vendor requirements

- Maintain information about storage topologies and configurations and ensure accuracy and availability to others

- Provide consulting services and support for release package testing, installation, deployment and operation

- Provide troubleshooting and technical support services for storage infrastructures

- Provide consulting and support services to identify storage operational, monitoring and reporting requirements

- Coordinate, schedule, implement and test storage tuning activities

- Coordinate and schedule repair services with 3rd party vendors for storage devices and validate that expected repairs and patches achieved expected benefits

- Label storage media components with identification tags and track media locations, and owners

- Plan, design, implement, test, operate and manage storage cache, storage types, channels, network, virtual pool and hierarchical storage configurations

- Manage and maintain inventories of spare media and equipment

- Provide media replication and validation

- Replace damaged storage media

- Process and configure storage services per provided requirements for data retention

- Reclaim unused storage

- Plan, architect and administer storage and tape virtualization solutions for storage assets

Service Initiation:

- Approved Work Requests

- Escalated Incidents or Problems from the Service Desk

Service Delivery Channels:

- Satisfied Work Requests

- Consulting and Support

Print Management

Description:

Provides a service to implement, install, maintain and support infrastructure printing assets used to underpin IT and business services.

Service Scope:

This service manages and maintains all infrastructure printing assets such as high speed printers, local printers, print queues, print spools, and virtual printers. It also includes other print related devices such as format plotters.

Service Functions and Features:

- Plan, install, configure and test print infrastructure components and connectivity

- Identify print capacity requirements needed to support services

- Identify physical facility requirements needed to operate printing components (i.e. floor space, equipment clearance, electrical, cooling, cabling, weight load)

- Provide print requirements to support procurement activities

- Decommission print infrastructure components upon request

- Maintain print components in compliance with supporting vendor requirements

- Maintain information about printer and print queue configurations and ensure accuracy and availability to others

- Provide consulting services and support for release package testing, installation, deployment and operation

- Provide troubleshooting and technical support services for print infrastructures

- Provide consulting and support services to identify printer and print queue operational, monitoring and reporting requirements

- Coordinate, schedule, implement and test print tuning activities

- Coordinate and schedule repair services with 3rd party vendors for print devices and validate that expected repairs and patches achieved expected benefits

- Label print device components with identification tags and track printer locations, and owners

- Plan, design, implement, test, operate and manage print queues and print queue configurations

- Plan design, implement, test and manage print spools

- Manage and maintain definitive hardware stores for printer spare parts and equipment

- Plan, architect and administer printer virtualization solutions for printing assets

Service Initiation:

- Approved Work Requests

- Escalated Incidents or Problems from the Service Desk

Service Delivery Channels:

- Satisfied Work Requests

- Consulting and Support

Fax Management

Description:

Provides a service to implement, install, maintain and support infrastructure faxing assets used to underpin business services.

Service Scope:

This service manages and maintains all infrastructure faxing assets such as networked fax devices and standalone fax devices. It also includes the management of fax templates, fax document libraries, fax phone numbers and fax broadcast lists.

Service Functions and Features:

- Plan, install, configure and test fax infrastructure components and connectivity

- Identify fax capacity requirements needed to support services

- Identify physical facility requirements needed to operate faxing components (i.e. floor space, equipment clearance, electrical, cooling, cabling, weight load)

- Provide fax requirements to support procurement activities

- Decommission fax infrastructure components upon request

- Maintain fax components in compliance with supporting vendor requirements

- Maintain information about fax configurations and ensure accuracy and availability to others

- Provide consulting services and support for release package testing, installation, deployment and operation

- Provide troubleshooting and technical support services for fax infrastructures

- Provide consulting and support services to identify fax operational, monitoring and reporting requirements

- Coordinate and schedule repair services with 3rd party vendors for fax devices and validate that expected repairs and patches achieved expected benefits

- Label fax device components with identification tags and track fax locations, and owners

- Plan, design, implement, test, operate and manage fax queues, fax queue configurations, fax templates, remote faxing, fax phone numbers and fax broadcast lists

- Test and validate fax connectivity and transmission

- Manage and maintain fax document libraries

- Manage and maintain definitive hardware stores for fax spare parts and equipment

Service Initiation:

- Approved Work Requests

- Escalated Incidents or Problems from the Service Desk

Service Delivery Channels:

- Satisfied Work Requests

- Consulting and Support

Physical Facilities Management

Description:

This service provides ongoing management and maintenance of physical site infrastructure assets used to house IT infrastructure hardware, supplies and people.

Service Scope:

This service manages and maintains all infrastructure physical site assets such as building core and shell, floor space, cabling, fire protection systems, mechanical, electrical, lighting, raised floor, desks, cubicles, console shelving and mounting.

Service Functions and Features:

- Provide consulting and expertise for physical site infrastructure assets such as floor space, mechanical, electrical, ventilation, cabling, lighting, raised floor, desks, cubicles, console shelving and mounting

- Provide consulting and expertise for local building and operating codes and zoning (safety) ordinances

- Manage and maintain physical site locations to provide clean operating environment free from litter, dust and pollutants

- Manage, test and maintain fire protection systems

- Monitor use of physical site premises power and utilities to identify upgrades or downgrades to accommodate changes in equipment capacity, local building codes, zoning, people or general physical infrastructure

- Oversee changes to physical site infrastructure to ensure existing services are not adversely impacted by physical site construction activities

- Ensure proper labeling of equipment, cables and telecommunication lines is in place and adequately maintained

- Oversee repairs to physical site infrastructure components done by 3rd parties and validate that repairs meet expected benefits

- Manage and schedule access to physical site conference and meeting rooms

- Manage and maintain conference and meeting room supplies, audio, video and teleconferencing equipment

Service Initiation:

- Approved Work Requests

- Escalated Incidents or Problems from the Service Desk

Service Delivery Channels:

- Satisfied Work Requests

- Consulting and Support

Telephony Management

Description:

This service provides operation, management and maintenance of telephony devices and assets.

Service Scope:

This service manages and maintains all infrastructure telephony assets such as desk phones, PBX, internet telephones (voice over IP), desk phones, cell phones, paging devices, pager services, dial plans, telephony service configurations and call center infrastructure.

Service Functions and Features:

- Plan, install, configure and test telephony hardware, applications and systems software configurations to meet desired calling functions and features

- Provide adequate voice quality and dial tone availability

- Process fulfillment requests for installation, moves, adds and changes to telephony equipment

- Configure telephony equipment for network access and perform testing to ensure equipment is operational and fit for purpose

- Identify physical facility requirements needed to operate telephony components (i.e. floor space, equipment clearance, electrical, cooling, cabling, weight load)

- Decommission telephony hardware, applications and related systems software upon request

- Provide telephony requirements to support procurement activities

- Obtain and manage telephone numbers to meet business needs

- Maintain telephony assets in compliance with supporting 3rd party vendor and lease requirements

- Maintain information about installed telephony hardware, application, software and networking configuration items and ensure accuracy and availability to others

- Label telephony equipment with asset tags and track locations, serial numbers and owners

- Provide consulting services and support for release package testing, installation, deployment and operation

- Provide troubleshooting and technical support services for telephony hardware, software and networking components

- Implement capacity planning and tuning actions for telephony assets

- Coordinate and schedule telephony repair services with 3rd party vendors and validate that expected repairs achieve expected benefits

- Maintain telephony Known Error and bug lists

- Provide consulting and support services to identify telephony operational, monitoring and reporting requirements

- Manage and maintain definitive hardware stores for telephony spare parts and equipment

Service Initiation:

- Approved Work Requests

- Escalated Incidents or Problems from the Service Desk

Service Delivery Channels:

- Satisfied Work Requests

- Consulting and Support

Personal Computing Device Management

Description:

This service operates, manages and maintains personal computing devices and related assets used by company employees.

Service Scope:

This service manages and maintains all infrastructure physical personal computing assets such as PCs, workstations, virtual PC images and laptops. It also manages related components such as device operating system software, device peripherals, locally attached storage devices and device cables.

Service Functions and Features:

- Plan, install, configure and test workstation hardware and operating systems software configurations

- Configure workstations for network access and perform connectivity testing to ensure desktops are recognized by network

- Identify physical facility requirements needed to operate workstations (i.e. equipment clearance, electrical, ventilation, etc.)

- Decommission and cleanse workstation hardware and related software and data upon request

- Provide workstation hardware, software and networking requirements to support procurement activities

- Install, modify or remove workstation hardware and software components on request

- Maintain workstation hardware and software assets in compliance with supporting 3rd party vendor and lease requirements

- Maintain information about installed workstation hardware, software and networking configuration items and ensure accuracy and availability to others

- Label workstations with asset tags and track workstation locations, serial numbers and owners

- Provide consulting services and support for release package testing, installation, deployment and operation

- Provide troubleshooting and technical support services for workstation hardware, software and networking components

- Implement tuning actions for workstation assets

- Coordinate and schedule workstation repair services with 3rd party vendors and validate that expected repairs and software patches achieved expected benefits

- Maintain Workstation Known Error and bug lists

- Manage and maintain definitive hardware stores for workstation spare parts and equipment

- Plan, architect and administer PC and PC operating system virtualization solutions for workstation assets

Service Initiation:

- Approved Work Requests

- Escalated Incidents or Problems from the Service Desk

Service Delivery Channels:

- Satisfied Work Requests

- Consulting and Support

Mobile Device Management

Description:

This service operates, manages and maintains employee (and employee owned bring your own device) mobile phones and tablets.

Service Scope:

This service manages and maintains all devices used to support mobile access to IT services. This is typically devices such as digital phones and tablets.

Service Functions and Features:

- Enforced Password management and control

- Device Wipe in the event of theft or loss

- Remote locking

- Configuration verification and validation

- Jailbreak and Rooted Detection enforcement and support

- Push/Pull of applications and configuration changes

- Support for apps store capabilities that allow users to self-load and install applications

- Audit Trail/Logging of device activity

- Support for standard industry operating systems

- Plan, install, configure and test device hardware and operating systems software configurations

- Configure devices for network access and perform connectivity testing to ensure devices are recognized by network

- Provide device hardware, software and networking requirements to support procurement activities

- Maintain device hardware and software assets in compliance with supporting 3rd party requirements

- Maintain information about device hardware, software and networking configuration items and ensure accuracy and availability to others

- Label devices with asset tags and track device serial numbers and owners

- Provide troubleshooting and technical support services for device hardware, software and mobility components

- Coordinate and schedule device repair services with 3rd party vendors and validate that expected repairs achieved expected benefits

- Maintain device Known Error and bug lists

- Manage and maintain definitive hardware stores for device equipment

Service Initiation:

- Approved Work Requests

- Escalated Incidents or Problems from the Service Desk

Service Delivery Channels:

- Satisfied Work Requests

- Consulting and Support

Specialized Device Management Services

Description:

This service operates, manages and maintains specialized devices and systems used to perform specific business functions that require IT related operations and support.

Service Scope:

This service manages and maintains specialized devices that require IT operations and support such as automated teller machines, lab equipment, copiers, scanners, shop floor control equipment, microwave towers and any other devices unique to the business.

Service Functions and Features:

- Plan, install, configure and test specialized device hardware and operating software configurations

- Configure specialized devices for network access and perform connectivity testing to ensure devices are recognized by network

- Identify physical facility requirements needed to operate devices (i.e. equipment clearance, electrical, ventilation, etc.)

- Decommission and cleanse device hardware and related software and data upon request

- Provide device hardware, software and networking requirements to support procurement activities

- Install, modify or remove device hardware and software components on request

- Maintain device hardware and software assets in compliance with supporting 3rd party vendor and lease requirements

- Maintain information about installed device hardware, software and networking configuration items and ensure accuracy and availability to others

- Label devices with asset tags and track device locations, serial numbers and owners

- Provide consulting services and support for release package testing, installation, deployment and operation

- Provide troubleshooting and technical support services for device hardware, software and networking components

- Implement tuning actions for device assets

- Coordinate and schedule device repair services with 3rd party vendors and validate that expected repairs and software patches achieved expected benefits

- Maintain device Known Error and bug lists

- Manage and maintain definitive hardware stores for device spare parts and equipment

Service Initiation:

- Approved Work Requests

- Escalated Incidents or Problems from the Service Desk

Service Delivery Channels:

- Satisfied Work Requests

- Consulting and Support

Virtualization Management

Description:

This service provisions and operates virtualization platforms that underlie servers, PCs and storage systems that are virtualized.

Service Scope:

Supplies managed by this service cover hardware and systems software platforms that virtualize servers, PCs, and storage systems. While this service does not manage those items directly, it does manage the underlying infrastructure that allows those items to be virtualized.

Service Functions and Features:

- Plan, architect and administer server virtualization solutions for server and operating system assets

- Configure and manage host configurations

- Provide and manage blueprint of validated virtual machine configurations—including networking, storage and security settings—and deploy it to hosts upon request

- Provide users with capabilities to manage their own subsets of the virtualized infrastructure (e.g. allow users to manage their own virtualization pools)

- Maintain security and restriction settings to manage user access to virtualized machines, resource pools and servers

- Provide minimum, maximum and proportional resource shares for CPU, memory, disk and network bandwidth

- Modify resource share allocations while virtual machines are running and enable applications to dynamically acquire more resources to accommodate peak performance needs

- Plan, install, configure and test virtualization hardware and operating systems software configurations

- Configure virtualization platforms for network access and perform connectivity testing to ensure platforms are recognized by network

- Identify physical facility requirements needed to operate virtualized physical hardware (i.e. floor space, equipment clearance, electrical, cooling, cabling, weight load)

- Decommission virtualization platforms upon request

- Provide virtualization hardware, software and networking requirements to support procurement activities

- Install or remove virtualized servers, PCs or storage virtual machines on request

- Maintain virtualization IP addresses, and physical locations

- Provide consulting services and support for virtualized platforms

- Provide troubleshooting and technical support services for virtualized platforms

- Implement capacity planning and tuning actions for virtual machines

- Coordinate and schedule server repair services with 3rd party vendors and validate that expected repairs and software patches achieved expected benefits

- Maintain virtualization platform Known Error and bug lists

- Provide consulting and support services to identify virtual machine operational, monitoring and reporting requirements

Service Initiation:

- Approved Work Requests

- Escalated Incidents or Problems from the Service Desk

Service Delivery Channels:

- Approved Work Requests

- Escalated Incidents or Problems from the Service Desk

IT Supplies Management

Description:

This service provisions and operates inventories of supplies needed to maintain and support IT services

Service Scope:

Supplies managed by this service might include items such as printer paper, printer toner, cleaning supplies, cleaning swabs and pads for equipment, blank media and spare media cases.

Service Functions and Features:

- Identify and manage catalog of supplies included within the scope of these services

- Requisition and fulfill orders for supplies upon demand or within scheduled intervals

- Monitor supply levels and re-order points and take actions to replenish supplies when levels fall below identified thresholds

- Coordinate warehousing and requisition of supplies with 3rd party suppliers

- Report on supply inventory levels and costs on a scheduled basis

Service Initiation:

- Approved Work Requests

- Escalated Incidents or Problems from the Service Desk

Service Delivery Channels:

- Satisfied Work Requests

- Consulting and Support

Middleware Transaction/Message Management

Description:

This service architects and manages transaction workflow and messaging middleware solutions that underpin services.

Service Scope:

This service manages assets that include transaction management middleware systems software such as message queues, messaging middleware and other types of connectivity software that control message queues or transaction flows.

Service Functions and Features:

- Plan, design, build, test, implement, manage and maintain transaction and messaging architectures and infrastructures to support needed IT services when requested

- Manage performance and capacity of transaction and message queues

- Ensure quality of service (QoS) for messages and transactions through control mechanisms such as queue managers and two/three phase commit solutions

- Establish and maintain transaction and messaging standards such as Queue Names, Queue Properties, indicators for Quality of Service, naming Services, Message Identification, message type (request, reply, report on exception) and others

95

- Provide consulting services to development groups for implementing and operating underlying transaction and message support for the solutions being built

- Establish, maintain and communicate API (Application Programming Interface) standards for accessing transaction and messaging services

- Configure and administer transactions and message queue managers including creation and deletion of queues, message routing, and dealing with different levels of message confirmations and acknowledgments.

Service Initiation:

- Approved Work Requests

- Escalated Incidents or Problems from the Service Desk

Service Delivery Channels:

- Satisfied Work Requests

- Consulting and Support

Chapter
5

Operational Management Support Services

This category includes services that manage and maintain IT operational workflow to support IT and business services across the physical infrastructure.

- Service Desk
- Service Monitoring
- Incident Response
- Problem Control
- Request Fulfillment
- Backup/Restore Management
- Job Schedule Management
- Dispatch and Break-Fix Support
- Clock Management
- Service Startup/Shutdown Management
- File Transfer and Control Management
- Archive Management
- Data Entry Support
- Report Packaging and Distribution Support

Service Desk

Description:

This service provides a single point of contact and communications for resolving incidents, fulfilling requests and dealing with a variety of service events.

Service Scope:

This service includes all IT and IT Business Support services that are running in production.

Service Functions and Features:

- Provide first call level investigation and diagnosis of reported incidents

- Escalate incidents and requests that cannot be resolved within agreed timescales or within Service Desk capabilities

- Coordinate resolution of incidents when reported to the Service Desk

- Plan, install, configure and maintain records of incident and request calls and categorize these for historical retrieval by others

- Communicate incident and request status to others

- Record, route and track service requests until they are fulfilled

- Identify estimates of needed labor headcount to operate the Service Desk

- Close incidents when resolved

- Provide consulting services and skills/labor estimates to support release package testing, installation, deployment and operation

- Conduct customer satisfaction call-backs or surveys to assess quality of provided services

- Plan, install and manage Service Desk operational function to provide local, centralized, virtual or follow-the-sun support goals

- Plan, install, configure and test call management hardware, software and networking configurations to manage calls to the Service Desk

- Maintain adequate levels of staffing skills to meet Service Desk goals and objectives

- Monitor and manage calling queues to make sure calls are handled in a timely manner

- Integrate Knowledge Management databases and repositories for call agent access

Service Initiation:

- Telephone contact

- Email request

- Web site self-help facility

- Automated alarm or alert

Service Delivery Channels:

- Telephone response

- Email response

- Web site self-help facility

- Formal Service Disruption Notice

- Consulting Support

Service Monitoring

Description:

This service captures events (IT alarms, alerts and notifications) in the IT infrastructure and forwards them to appropriate personnel or systems for further action.

Service Scope:

This service covers all infrastructure components and services in production.

Service Functions and Features:

- Collect and coordinate IT infrastructure events and event triggers to be acted upon

- Prioritize, escalate and forward events to personnel or systems to be acted upon

- Provide filtering and event correlation mechanisms to reduce event noise and ensure events are forwarded to their root sources for action

- Broadcast and display events to other parties and systems that may need to be aware of them upon request

- Provide event logging and historical repositories to aid in the investigation of incidents, problems and overall service quality

- Plan, design, build, test, implement event management toolsets and manage automated responses to events to prevent or reduce service outages and unplanned labor

- Maintain event tables and rules in accordance with changes to the IT infrastructure

- Process requests for event monitoring and handling in accordance with a consistent request fulfillment process

- Design, build, implement, test and maintain console solutions to consolidate views of service events or highlight them for support staff

Service Initiation:

- Approved work requests

- Escalated incidents and problems from the Service Desk

Service Delivery Channels:

- Infrastructure service alarm and alert notifications to operator console, monitoring console, pager, email or other target communication device

- Consulting support

Incident Response

Description:

This service coordinates actions to restore normal business operations as quickly as possible and minimizes the adverse impact on business operations when agreed service levels are threatened.

Service Scope:

This service includes all IT and IT Business Support services that are running in production.

Service Functions and Features:

- Identify and manage channels for notification of incidents such as escalated Service Desk calls, automated events that have been forwarded, emails, web interfaces and direct telephone calls to support staff

- Ensure all reported incidents are appropriately logged and adequately described

- Define, maintain and manage a standard incident schema to categorize all incidents when received

- Define, maintain and manage a standard set of priority classification criteria that will quickly prioritize incidents when received

- Triage and escalate incidents to appropriate support staff or 3rd party suppliers for resolution

- Coordinate all activities to investigate, diagnose, recover and resolve incidents across the IT infrastructure

- Coordinate activities to ensure incidents and their status are appropriately communicated to those that are impacted by incidents or need to be aware of them

- Provide follow-up to ensure that incidents are appropriately resolved, closed and documented

Service Initiation:

- Approved work requests

- Escalated incidents from the Service Desk

Service Delivery Channels:

- Resolved incidents escalated from the Service Desk

- Consulting support

Problem Control

Description:

This service proactively reduces incident occurrences by identifying their root causes and takes actions to coordinate development of workarounds for Known Errors and actions to remove those errors from the IT infrastructure.

Service Scope:

This service includes all IT and IT Business Support services that are running in production.

Service Functions and Features:

- Identify and manage channels for notification of problems such as escalated Service Desk calls, automated events that have been forwarded, emails, web interfaces and direct telephone calls to support staff

- Ensure all reported problems are appropriately logged and adequately described

- Define, maintain and manage a standard problem schema to categorize all problems when received

- Define, maintain and manage a standard set of priority classification criteria that will quickly prioritize problems when received

- Triage and escalate problems to appropriate support staff or 3rd party suppliers for resolution

- Coordinate all activities to investigate, diagnose and resolve problems across the IT infrastructure

- Coordinate activities to ensure problems and their status are appropriately communicated to those that are impacted by problems or need to be aware of them

- Provide follow-up to ensure that problems are appropriately resolved, closed and documented

- Proactively review incidents to identify incident trends that could be reversed by removing their root cause

- Identify, manage and maintain a list of infrastructure Known Errors from the IT infrastructure, development and transition staff for new services with workarounds and ensure these are communicated to the Service Desk and appropriate personnel

- Provide service disruption notices and communications to appropriate personnel and senior management staff for major incidents that have occurred

- Define, maintain and manage an inventory of problem solving techniques and provide training and communications on these to support staff

Service Initiation:

- Approved work requests

- Escalated incidents from the Service Desk

Service Delivery Channels:

- Resolved incidents escalated from the Service Desk

- Consulting support

- Completed work requests

- Problem status reporting

Request Fulfillment

Description:

This service provides a standard single point of contact channel for company employees to request and receive standard IT support services and information.

Service Scope:

This service includes all IT and IT Business Support services that are running in production.

Service Functions and Features:

- Plan, design, build, test, implement and manage automated interfaces for placing service requests such as an ordering web site or Service Desk interface

- Identify and manage catalog of standard services

- Requisition and fulfill orders for services upon demand or within scheduled intervals

- Provide pre-defined channels and procedures for receiving and fulfilling each service identified in the catalog

- Fulfill orders for standard services within pre-defined time limits

- Ensure requests for standard services are appropriately authorized and financed

- Coordinate requisition and fulfillment activities with 3rd party vendors as needed

- Log all requests and maintain a history of requests received and their status

- Coordinate all activities to ensure requests are fulfilled to the satisfaction of customers and clients and ensure they are closed appropriately

- Coordinate fulfillment of requests with appropriate change, configuration and asset control services

Service Initiation:

- Approved work requests

- Escalated requests from the Service Desk

Service Delivery Channels:

- Completed work requests

- Request status notifications

- Resolved requests that were escalated from the Service Desk

Backup/Restore Management

Description:

This service provides backup and restoration of data, applications and systems used to underpin IT and IT Business Support services.

Service Scope:

This service includes all infrastructure data, systems and applications used to support IT development and production services.

Service Functions and Features:

- Process requests for backup services by translating backup and restore requirements to operational solutions that provide needed backup and restore operations

- Plan, design, build, implement, test and manage backup and restore procedures per requested requirements

- Build and maintain run books describing backup and restore procedures

- Design, build, test and maintain backup and restore scripts and processes

- Perform backups per identified schedules

- Execute backup activities, scripts and processes

- Execute restore activities, scripts and processes when requested

- Monitor and report backup/restore progress

- Monitor backups to ensure completeness and accuracy

- Maintain report logs of backups taken and backup success status

- Cycle backups per established policies

- Maintain media rotation cycles for backups

Service Initiation:

- Approved work requests

- Incidents and requests escalated from the Service Desk

- Approved operational run procedures

Service Delivery Channels:

- Completed work requests

- Backup and restore status notifications

- Resolved incidents and requests that were escalated from the Service Desk

- Completed operational run procedures

- Consulting support

Job Schedule Management

Description:

This service schedules and operates planned jobs and events and ensures they have completed successfully within agreed timeframes.

Service Scope:

This service includes all IT and IT Business Support services that are running in production.

Service Functions and Features:

- Process requests for scheduling services by translating schedule requirements to operational solutions that provide needed scheduling operations

- Schedule batch jobs and events for processing

- Plan, design, build, test, implement and maintain job and event schedules

- Coordinating and maintain multiple jobs and events within single streams of work to ensure proper sequencing of job tasks

- Develop schedule restart and recovery solutions in the event schedules cannot complete as planned

- Develop scenarios and methods to re-start failed or stopped jobs and events

- Log job and event restarts and failure causes and interface schedule incidents with incident management services and activities

- Perform job and event runtime trend analysis to proactively identify future threats to successful schedule completion

- Monitor schedules for backups

Service Initiation:

- Approved work requests

- Incidents and requests escalated from the Service Desk

- Approved operational run procedures

Service Delivery Channels:

- Completed work requests

- Schedule status notifications

- Resolved incidents and requests that were escalated from the Service Desk

- Completed operational run procedures

- Consulting support

Dispatch and Break-Fix Support

Description:

This service dispatches field support personnel to provide onsite touch services for troubleshooting, repair and installation of IT assets at physical site locations.

Service Scope:

This service covers all IT assets physically located at customer sites as well as the resource staff to install, repair and troubleshoot them. Examples can include printers, servers, workstations, and networking devices.

Service Functions and Features:

- On-site hands on troubleshooting, repair and installation assistance at the customer sites

- On-site replacement of parts

- Management of spare parts inventories and replacement equipment where needed

- Installation and configuration of software where it cannot be done remotely by any other means

- Coordination with asset warranty requirements and constraints

- Provision of touch services to maintain IT assets

- Troubleshooting of network and internet connectivity checking for physical connection issues and failures

- Management of dispatch personnel, their work schedules and locations

- Dispatch of work personnel to onsite locations within established service targets and policies

- Coordination with local site access and security procedures allowing dispatch personnel with physical access to on-site equipment

- Unpacking of containers and packages that local site equipment may have been shipped with

- Removal and disposal of equipment, containers, boxes or other materials from sites

- Transfer of equipment from local site loading dock to equipment locations

- Updating of work tickets and work requests with status of equipment repair, installation and dispatched personnel

Service Initiation:

- Approved work requests

- Incidents and requests escalated or dispatched from the Service Desk

- Pager or other device notifications

Service Delivery Channels:

- Completed work requests

- Schedule status notifications

Clock Management

Description:

This service provides coordination of system clocks and timing services across the IT infrastructure.

Service Scope:

This service includes timing and clock infrastructures that underpin all IT and IT Business Support services that are running in development and production.

Service Functions and Features:

- Define policies, plans and strategies to synchronize time clocks across the IT infrastructure

- Identify and maintain authorized sources for clock times and dates

- Plan, design, build, test, implement and maintain local and global time synchronization solutions across the IT infrastructure

- Plan, design, build, test, implement and maintain strategies to handle unique clock events such as daylight savings, leap year and local time variances

- Identify and apply a definitive set of standard sources for synchronizing clocks (i.e. Internet NTP)

- Verify daylight savings impacts/changes

Service Initiation:

- Approved work requests

- Incidents and requests escalated from the Service Desk

- Approved operational run procedures

Service Delivery Channels:

- Completed work requests

- Clock and timing APIs for applications

- Resolved incidents and requests that were escalated from the Service Desk

- Completed operational run procedures

- Consulting support

Service Startup/Shutdown Management

Description:

This service provides coordination and synchronization of pre-requisite, co-requisite and post requisite activities for starting and shutting down IT and IT Business Support services.

Service Scope:

This service includes all IT and IT Business Support services that are running in production.

Service Functions and Features:

- Define, plan, build, test, implement and maintain service startup procedures

- Define, plan, build, test, implement and maintain service shutdown or suspension procedures

- Coordinate activities across the IT infrastructure to successfully startup, shutdown or suspend services whether planned, scheduled or requested

- Document startup/shutdown/suspension work flows and procedures

- Identify and implement strategies to automate the startup or shutdown of services minimizing manual intervention where possible

- Monitor startup/shutdown processes to ensure they have completed successfully and within desired schedules and timeframes

- Record the total amount of time required for services to be started and brought to fully operational state

- Maintain criteria that ensures and validates that started services are fully operational and available for use

- Log all incidents related to startup and shutdown activities

Service Initiation:

- Approved work requests

- Incidents and requests escalated from the Service Desk

- Approved operational run procedures

Service Delivery Channels:

- Completed work requests

- Resolved incidents and requests that were escalated from the Service Desk

- Completed operational run procedures

- Consulting support

File Transfer and Control Management

Description:

This service provides coordination and synchronization of pre-requisite, co-requisite and post requisite activities for transferring files across the infrastructure or to external parties.

Service Scope:

This service includes all IT and IT Business Support services that are running in production.

Service Functions and Features:

- Define, plan, build, test, implement and maintain scripts, applications and jobs to execute file transfer activities

- Coordinate activities to execute initiation of file transfers

- Coordinate activities to schedule and automate execution of file transfers

- Validate that appropriate security controls are not compromised during transfer execution activities

- Execute any pre and post transfer processing such as encrypting, massaging and decrypting files

- Log and maintain history of all transfer activity

- Validate that transfer activities took place successfully and within agreed timeframes and schedules

- Ensure successful and secure transfer via logging function

- Define, plan, build, test, implement and maintain scripts, applications and jobs to restart or recover file transfer activities in the event of unsuccessful transfers

Service Initiation:

- Approved work requests

- Incidents and requests escalated from the Service Desk

- Approved operational run procedures

Service Delivery Channels:

- Completed work requests

- Resolved incidents and requests that were escalated from the Service Desk

- Completed operational run procedures

- Consulting support

Archive Management

Description:

This service archives or retrieves infrastructure data, documents, applications and systems on behalf of development and production services.

Service Scope:

This service includes all data, applications and systems that underpin all IT and IT Business Support services.

Service Functions and Features:

- Implement and maintain an inventory of data to be archived in accordance with business policy

- Implement, manage and maintain pick lists of files to go to archive facilities

- Coordinate activities to determine and agree media to be used for archiving (e.g. hardcopy print, microfiche, tape, CD-ROM, etc.)

- Coordinate activities to determine the physical media type and location for archived media

- Identify criteria and requirements for archive physical locations (e.g. protection against humidity, theft, fire, etc.)

- Coordinate activities to determine and agree schedules and frequency for archiving

- Periodically test storage and retrieval of items from archive and validate that media is usable

- Log and maintain historical records of archiving activities

- Maintain status lists showing which items are in archives versus other locations (such as for retrieved archive items)

- Maintain accurate and current pick lists for archiving in accordance with changes to business policies and needs

- Take actions to retrieve items from archives or store them there per schedules and requests

- Remove media from archives that is no longer needed

- Develop and maintain archive schedules

- Coordinate archive activities with Service Continuity Management services

Service Initiation:

- Approved work requests

- Incidents and requests escalated from the Service Desk

- Approved operational run procedures

Service Delivery Channels:

- Completed work requests

- Resolved incidents and requests that were escalated from the Service Desk

- Completed operational run procedures

- Consulting support

Data Entry Support

Description:

This service records and registers input data from external sources into the IT infrastructure and makes it available for use by services.

Service Scope:

This service includes all IT and IT Business Support services running in production.

Service Functions and Features:

- Plan, design, build, implement and test strategies and technologies for inputting external data into the IT infrastructure

- Identify labor requirements such as skills and headcount for executing data entry activities

- Coordinate activities to implement and manage technologies used to input data such as data entry terminals, devices or key-to-disk solutions

- Monitor and manage data entry activities to ensure planned volumes of data are input successfully

- Validate entered data to ensure it is accurate and in a format usable for services

- Schedule and monitor levels of labor headcount to align with planned increases or decreases in data volumes based on business activities

Service Initiation:

- Approved work requests

- Incidents and requests escalated from the Service Desk

- Approved operational run procedures

Service Delivery Channels:

- Completed work requests

- Resolved incidents and requests that were escalated from the Service Desk

- Completed operational run procedures

- Consulting support

Report packaging and Distribution Support

Description:

This service replicates reports and distributes them to company employees, management and customers.

Service Scope:

This service includes all IT and IT Business Support services running in production. It also includes any production activities involved with packaging, binding and collating reports as well as preparation of report media.

Service Functions and Features:

- Build and maintain a catalog of reports for production and distribution

- Build and maintain report distribution lists and schedules

- Plan, design, build, implement, test and maintain strategies and technologies for producing and distributing reports

- Provide report production services to print, collate, bundle, bind and package reports

- Provide report distribution services to deliver reports either via mail, website, email attachment or general file transfer

- Identify labor requirements such as skills and headcount for executing report distribution activities

- Validate that reports are distributed successfully and resolve any delivery complaints and issues

Service Initiation:

- Approved work requests

- Incidents and requests escalated from the Service Desk

- Approved operational run procedures

Service Delivery Channels:

- Completed work requests

- Resolved incidents and requests that were escalated from the Service Desk

- Completed operational run procedures

- Consulting support

Chapter

6

Service Transition Support Services

This category includes services that transition new or changed service solutions to production operations.

- Release Planning and Packaging
- Service Deployment and Decommission
- Site Preparation Support
- Service Validation and Testing Support
- Training Support
- Organizational Change Support
- Knowledge Management

Release Planning and Packaging

Description:

This service plans, schedules and coordinates service and infrastructure changes into release packages.

Service Scope:

This service includes all IT and IT Business Support services targeted for production operations.

Service Functions and Features:

- Identify strategies for deploying releases (e.g. "big-bang", parallel, phased releases, etc.)

- Identify scope and configuration items for major, minor, and emergency release packages

- Build release packages

- Identify and develop service and release design and acceptance criteria

- Identify requirements for master copies of media to deploy release packages

- Build, test and certify master definitive copies of media to deploy release packages

- Plan, arrange and coordinate temporary additional or swing support facilities and people needed to transition services or initially operate them

- Develop and administer the company release policy

- Control versions and version numbering for releases

- Identify strategies for rolling back or recovering releases in the event of failures

- Support Post Implementation Review sessions when requested by Change Control Services

Service Initiation:

- Approved work requests

- Incidents and requests escalated from the Service Desk

Service Delivery Channels:

- Completed work requests

- Resolved incidents and requests that were escalated from the Service Desk

- Published plans and strategies

- Consulting support

Service Deployment and Decommission

Description:

This service plans, manages and coordinates activities to deploy or decommission service solutions as set out in release packages.

Service Scope:

This service includes all IT and IT Business Support services running in production.

Service Functions and Features:

- Assess readiness for deployment/decommission

- Control and gate release components so that only authorized and approved components are moved to production

- Develop and manage service transition/service decommission plans

- Deploy service supporting infrastructure, data and applications

- Schedule, monitor and control deployment activities and well as report on deployment status

- Manage replication and distribution of software and media components

- Coordinate activities to transition service supporting organization and people

- Decommission, retire and redeploy service assets as requested

- Validate completion of deployment and decommission activities

- Coordinate approvals to execute deployment/decommission plans

- Coordinate deployment/decommission of service contracts, equipment leases, and working capital

- Design, build, test and implement prerequisite and post-requisite deployment scripts as needed

Service Initiation:

- Approved work requests

Service Delivery Channels:

- Completed work requests

- Published plans and strategies

- Consulting support

Site Preparation Support

Description:

This service manages and coordinates activities to prepare physical sites for new or changed infrastructure that will house service assets to support IT and IT Business Support services.

Service Scope:

This service includes all IT and IT Business Support services running in production or targeted for production.

Service Functions and Features:

- Identify site requirements such as floor space, electrical, mechanical, raised floor, ventilation, door clearances for moving equipment, equipment clearances, floor weight loads, desks, cubicles, console shelving and mounting

- Ensure requirements compliance with local building codes, safety and zoning

- Build floor space plans to accommodate expected IT equipment, power, people, furniture, environmental equipment, ventilation, cabling and physical security needs

- Provide consultation support for site selection, selection of 3rd party site contractors and contracting for site preparation services from 3rd parties

- Develop, administer and conduct site surveys

- Validate site physical space capacity, appropriate power, cabling and adherence to local standards and codes

- Plan, schedule, coordinate and manage physical site preparation activities

- Certify sites for readiness to accept service infrastructure and completion of 3rd party contractor obligations

- Provide consulting services to build and oversee plans to move, consolidate, expand or replicate physical data centers and operational sites

- Provide consulting services to plan and oversee specialized physical infrastructure projects such as adding a command or operational center within an existing building

- Provide cost estimates for proposed physical site build, consolidation or expansion projects

- Coordinate site and construction activities for unique equipment needs such as building Microwave Towers

Service Initiation:

- Approved work requests

Service Delivery Channels:

- Completed work requests

- Resolved requests and incidents escalated from the Service Desk

- Published plans and strategies

- Published site assessments

- Site blueprints

- Published contractor and construction specifications

- Consulting support

Service Validation and Testing Support

Description:

This service ensures that new or changed IT services and their supporting infrastructure are adequately tested and accepted for successful delivery and operation.

Service Scope:

This service includes all IT and IT Business Support services under development that are targeted for production.

Service Functions and Features:

- Certify new or changed services for production operations

- Develop and manage test plans

- Develop testing scripts

- Manage and oversee testing activities

- Provide repositories and sources for generic test data

- Manage, configure and administer test lab facilities

- Schedule testing activities and use of test lab facilities

- Provide consulting expertise for development of testing strategies, test conditions, test cases, test cycles, service acceptance criteria and expected results

- Provide independent validation of test results and certify that services meet expected acceptance criteria

- Plan, coordinate and manage service rehearsal and pilot testing activities

Service Initiation:

- Approved work requests

Service Delivery Channels:

- Completed work requests

- Published testing plans and strategies

- Test data repositories

- Published test results

- Published test conditions and expected results

- Consulting support

Training Support

Description:

This service develops training content and manages training activities needed to operate, support or use an IT service or IT supported business service successfully.

Service Scope:

This service includes all IT and IT Business Support services under development that are targeted for production as well as all services already in production.

Service Functions and Features:

- Design, develop, build and test training content and materials

- Identify requirements and coordinate activities to build and implement infrastructures used to deliver training

- Coordinate and schedule training sessions and classes

- Monitor personnel for completion of training and 3rd party certifications

- Coordinate activities with 3rd party training providers when requested

- Provide time and cost estimates for building, production of training content and delivering training

- Coordinate activities to store and administer distribution and warehousing of training content

Service Initiation:

- Approved work requests

Service Delivery Channels:

- Completed work requests

- Classroom sessions

- Training presentations

- Online training sessions

- Published training content

- Consulting support

Organizational Change Support

Description:

This service ensures that the appropriate organization, skills and employee acceptance is in place to operate services successfully.

Service Scope:

This service includes all IT and IT Business Support services under development that are targeted for production as well as all services already in production that are undergoing large changes. It targets the stakeholders (Users, support staff, management) for acceptance of new services and service changes.

Service Functions and Features:

- Provide consulting services that identify the appropriate organizational roles, responsibilities and structure are in place to operate services successfully

- Identify strategies and conduct activities to reduce resistance and organizational barriers to new or changed services

- Conduct organizational assessments to identify existing gaps in skills, organization, organizational impact, culture and readiness towards change to operate services successfully

- Develop communication strategies and plans for new or changed services

- Monitor the organizational acceptance of new services and service changes while they are being transitioned

- Provide oversight for execution of communication plans including training programs

Service Initiation:

- Approved work requests

Service Delivery Channels:

- Completed work requests

- Communication plans and strategies

- Service communication presentations

- Communication meetings

- Newsletters

- Email campaigns

- Organizational assessment reports

- Consulting support

Knowledge Management

Description:

This service manages and administers knowledge repositories, subscriptions and information to ensure that the right information is delivered to the appropriate place or person at the right time.

Service Scope:

This service includes all IT and IT Business Support services in production or targeted for production.

Service Functions and Features:

- Administer and manage documentation subscriptions

- Design, create and maintain operational run books

- Maintain and manage document libraries and repositories

- Plan, design, build, test, implement and maintain software and hardware solutions for managing document and information repositories and websites

- Design, create and maintain templates to be used for documenting operational tasks and infrastructure architecture

- Maintain and manage knowledge databases, repositories and web portals

- Customize and administer Knowledge Management tools

143

- Provide consulting services for documenting new services, processes, functions and procedures and assessing documentation quality

- Coordinate activities to ensure Knowledge is appropriately secured, current and available in the case of a disaster

- Support service availability improvements by proactively identifying where Knowledge can improve availability, reduce outage times or labor costs

- Remove, archive and decommission documentation and knowledge no longer needed

Service Initiation:

- Approved work requests

- Requests escalated from the Service Desk

- Approved operational run procedures

Service Delivery Channels:

- Completed work requests

- Completed requests escalated from the Service Desk

- Knowledge repositories

- Online knowledge systems

- Consulting support

Chapter
7

Service Design and Build Services

This category includes services that plan, build and construct new or changed service solutions.

- Operational Planning and Consulting
- Solution Planning and Development
- Development Support Operations
- Capacity Management
- Availability Management
- Service Continuity Management
- Website Support

Operational Planning and Consulting

Description:

This service provides consulting and expertise to develop the plans and strategies for the operational infrastructure that underpin new and changed solutions. This expertise serves to protect the IT investment in building solutions by making sure they are deployable and operable.

Service Scope:

This service includes all IT and IT Business Support services in production that are undergoing change new services that are targeted for production.

Service Functions and Features:

- Provide consulting services for operational solutions for each stage of the systems development lifecycle working closely with application solution development teams to ensure that appropriate levels of operability will be built into their solutions

- Provide support for building business cases for proposed new or changed services

- Develop Operational Readiness Plans

- Develop strategies for operational solutions to support services

- Provide operational consulting expertise upon request

- Design, build, test and deploy operational support tools when requested

- Certify solutions as ready for production operations

Service Initiation:

- Approved work requests

- Requests escalated from the Service Desk

Service Delivery Channels:

- Completed work requests

- Completed requests escalated from the Service Desk

- Consulting support

Solution Planning and Development

Description:

This service plans, designs, builds and tests applications to support business services, processes or functions.

Service Scope:

This service includes all IT and IT Business Support services in production that are undergoing change and new services that are targeted for production.

Service Functions and Features:

- Develop plans and strategies for new or existing applications

- Develop designs for applications to support services

- Build and integrate application solution components

- Conduct fit for purpose testing for applications

- Provide assistance and consulting support for transitioning applications into production

- Validate successful functionality of applications once implemented

- Identify and communicate application Known Errors that will be going into production

Service Initiation:

- Approved work requests

- Approved project charter

Service Delivery Channels:

- Completed work requests

- Completed projects

- Consulting support

Development Support Operations

Description:

This service plans, designs, builds, tests and operates the operational infrastructure to support applications development activities.

Service Scope:

This service includes all IT and IT Business Support services undergoing development activities.

Service Functions and Features:

- Develop plans and strategies for implementing and operating the operational infrastructure to support an application development effort

- Develop Operational Level Agreements to support application development efforts based on their needs and requirements

- Install and customize hardware and systems software to meet the needs of application development efforts

- Manage and operate the operational infrastructure to support the needs of application development efforts

- Migrate application components between development, unit, systems and quality assurance test environments

- Implement and customize development support tools such as compilers, test tools, database systems software and software configuration tools

- Backup and restore software code, databases and data upon request by development teams

Service Initiation:

- Approved work requests
- Approved project charter

Service Delivery Channels:

- Completed work requests

- Completed projects

- Consulting support

Capacity Management

Description:

This service ensures that all current and future capacity and performance aspects of the IT infrastructure are provided to meet business and service requirements at acceptable cost.

Service Scope:

This service includes all IT and IT Business Support services in production that are undergoing change, new services that are targeted for production and existing services already in production.

Service Functions and Features:

- Control access to services via demand management for business, service and resource capacity activities

- Develop capacity models for business, service and resource capacity activities

- Translate business forecasts into service and resource impacts and forecasts through workload characterization

- Provide application sizing estimates for business, service and resource capacity activities

- Provide capacity plans for business, service and resource capacity activities

- Perform capacity monitoring, analysis and provide recommendations for tuning activities

- ·Coordinate and implement capacity-related changes

- Provide management information about Capacity Management quality and operations

Service Initiation:

- Approved work requests

- Requests, incidents and problems escalated from the Service Desk

- Approved operational procedures

Service Delivery Channels:

- Completed work requests

- Completed and resolved requests, incidents and problems that are escalated from the Service Desk

- Published capacity reports and projections

- Consulting support

Availability Management

Description:

This service optimizes the capability of the IT infrastructure, services and supporting organization to deliver sustained levels of service availability that meet business requirements at acceptable costs.

Service Scope:

This service includes all IT and IT Business Support services in production that are undergoing change, new services that are targeted for production and existing services already in production.

Service Functions and Features:

- Provide consulting services to identify redundancy, failover and work-around requirements to meet service targets and business needs

- Provide consulting services to identify service improvement actions and strategies

- Provide operational and service risk assessments

- Develop and maintain availability plans and strategies

- Identify current capabilities and gaps in meeting proposed and existing service levels and targets

- Coordinate and lead service improvement activities and projects

- Provide requirements for monitoring service availability and track overall availability of services

Service Initiation:

- Approved work requests

- Requests, incidents and problems escalated from the Service Desk

Service Delivery Channels:

- Completed work requests

- Completed and resolved requests, incidents and problems that are escalated from the Service Desk

- Published availability plans and strategies

- Service availability risk assessments

- Consulting support

Service Continuity Management

Description:

This service supports business continuity management functions by ensuring that IT services can be recovered in the event of a major business disruption within required timescales.

Service Scope:

These services include all IT and IT Business Support services running in production deemed vital to business operations.

Service Functions and Features:

- Identify service continuity requirements based on business continuity plans

- Coordinate classification and priority of recovery events and activities based on provided priorities from the business

- Conduct service continuity risk assessments

- Develop service continuity recovery strategies and approaches

- Develop and maintain service continuity plans

- Test service continuity plans

- Implement and coordinate training and communications for service continuity plans, actions, roles, responsibilities and overall awareness

- Conduct reviews and audits of service continuity plans and report results to management

- Provide information on costs and cost alternatives for continuity strategies

- Review proposed and pending infrastructure changes to ensure they do not compromise continuity plans and strategies

Service Initiation:

- Approved work requests

- Formal invocation request from the Service Desk or authorized IT executive

- Approved continuity test schedules and operational procedures

Service Delivery Channels:

- Completed work requests

- Offsite operational location

- Published service continuity plans and strategies

- Service continuity risk assessments

- Consulting support

Website Support

Description:

This service provides plans, testing, construction and maintenance of internet web sites and pages that display company content.

Service Scope:

This service provides support for all company internet and intranet websites used in production.

Service Functions and Features:

- Create web site designs, graphics and visuals for displaying provided content

- Plan, build, maintain and manage corporate standards for web site design templates, colors, font sizes, company logo and GUI requirements

- Coordinate and lead activities to construct web pages and web sites

- Coordinate and lead activities to implement E-Commerce support if requested

- Coordinate domain name registrations

- Obtain, register and manage IP addresses

- Register web sites with internet search engines when requested

- Coordinate activities to manage, control and approve web site content for use and appropriateness

- Coordinate and manage web hosting services

Service Initiation:

- Approved work requests

- Requests and incidents escalated from the Service Desk

- Approved project charter

Service Delivery Channels:

- Completed work requests

- Completed and resolved requests and incidents escalated from the Service Desk

- Completed projects

- Consulting support

Chapter

8

Strategy and Control Services

This category includes services that govern and manage how services are delivered and how quality will be maintained.

- IT Service Strategy Support
- Architecture Management and Research
- IT Financial Management
- IT Project Management
- Change Control
- Configuration and Asset Management
- Lease and License Management
- Access and Security Management
- Service Audit and Reporting
- IT Workforce Management
- Procurement Support
- Process Management
- Supplier Relationship Management

IT Service Strategy Support

Description:

This service provides oversight and control over IT services and how they are to be delivered and supported both currently and in the future.

Service Scope:

All IT services in production, development or proposed.

Service Functions and Features:

- Define, build and maintain a catalog of available services

- Conduct periodic assessments to determine how well services are aligned with and meeting business objectives, whether some services should be decommissioned or new services added

- Coordinate activities to identify how services will be sourced and delivered

- Coordinate, schedule and manage meetings of the Service Governing Board

- Define, document and agree Service Level Agreements for each provided service that identify the services to be provided and their service targets

- Define, document and agree Operational Level Agreements used to support each service

- Provide information on services and service targets to be monitored along with their thresholds

- Initiate service improvement actions and coordinate service delivery and support issues to resolution

- Manage relationships with the business by proactively seeking out service needs, changes to service targets and assessing overall satisfaction with the services being delivered

- Provide coordination and escalation for exceptions to services that need to be addressed

- Conduct regularly scheduled reviews of service quality and ability to meet desired service targets

Service Initiation:

- Approved work requests

- Requests escalated from the Service Desk

- Requests escalated from IT Business Liaisons

Service Delivery Channels:

- Service portfolio and catalog

- Published IT service strategies

- Published service recommendations

- Completed work requests

- Completed and resolved requests escalated from the Service Desk and Business Liaisons

- Consulting support

Architecture Management and Research

Description:

This service controls the viability of the service infrastructure architecture and provides services that optimize and control that infrastructure through standards, controls and supporting organization.

Service Scope:

All IT services in production, development or proposed.

Service Functions and Features:

- Review operational day-to-day labor activities and proactively identify opportunities for automation to reduce manual labor efforts

- Maintain and manage and control application development and infrastructure standards

- Process, escalate, review and coordinate approvals for requests for changes to infrastructure standards

- Communicate approved architectures available for reuse

- Conduct ongoing research of new infrastructure solutions, techniques and tools and identify candidates to support current and future business needs

- Coordinate tool selection efforts when requested

- Coordinate activities to evaluate and approve architecture changes

- Manage and maintain repositories of reusable application components

- Coordinate ownership, development and maintenance over infrastructure service support and delivery toolsets that are shared across different services

Service Initiation:

- Approved work requests

- Requests escalated from the Service Desk

- Formal requests for architecture exceptions

Service Delivery Channels:

- Published architecture standards

- Architecture plans and strategies

- Architecture assessments

- Published architecture recommendations

- Approved or rejected architecture exceptions

- Published architecture research

- Completed work requests

- Completed and resolved requests escalated from the Service Desk

- Consulting support

IT Financial Management

Description:

This service provides IT accounting, budgeting and chargeback services for the infrastructure that supports IT and IT supported business services.

Service Scope:

. All IT services in production, development or proposed.

Service Functions and Features:

- Build and maintain the IT infrastructure budget

- Develop cost models and estimates for new or changes services

- Maintain accounting ledgers for services

- Interface financial information with business corporate financial practices and requirements

- Provide reports and information on service costs, budgets and revenue to others based on reporting schedules and ad-hoc requests

- Plan, design, build, test, implement and maintain IT financial chargeback systems and solutions

- Agree cost allocations, chargeback algorithms and assumptions used to charge for services (may not be in scope for ESM)

- Prepare and issue bills and payment/recovery status reports for IT chargeback activities (may not be in scope for ESM other than preparation of usage reports)

- Support internal and external corporate audit activities by providing IT financial information when requested

Service Initiation:

- Approved work requests

- Approved operational run procedures

Service Delivery Channels:

- Published IT budgets

- Published IT financial reports

- Published service cost models and estimates

- Service chargeback bills

- Completed work requests

- Consulting support

IT Project Management

Description:

This service provides project management oversight and project management services for IT projects to make sure they are completed on time, within budget and within scope.

Service Scope:

All IT services in development or proposed.

Service Functions and Features:

- Manage and maintain the portfolio of IT projects

- Provide project management office controls and oversight for IT programs

- Manage projects from inception through completion ensuring all planned work products are completed on time, within schedule and within budget

- Escalate project issues and coordinate resolution actions

- Prepare and communicate project plans, project status and budget reports

- Ensure project scope is tightly controlled and process changes to scope through change orders and Change Control services

- Coordinate activities to staff projects to meet business objectives

- Coordinate activities to obtain appropriate approvals and signoffs for project work products

- Manage variances to project plans and activities

- Provide administrative activities and coordination to close projects

Service Initiation:

- Approved work requests

- Approved project charter

Service Delivery Channels:

- Completed work requests

- Completed projects

- Project status and management reports

- Consulting support

Change Control

Description:

This service provides controls that protect availability and quality of services while changes are made to the IT infrastructure.

Service Scope:

All IT services in production, development or proposed.

Service Functions and Features:

- Record and classify changes as they are requested

- Publish and communicate which changes have been requested as well as status and implementation success or failure of changes

- Coordinate activities to agree and approve or reject requested changes on a timely basis to meet business needs

- Coordinate activities to identify service and technical impacts associated with requested changes

- Coordinate activities to validate change success and close changes

- Conduct and hold post implementation reviews for changes that meet specified criteria to require them

- Track and report on progress of changes as they proceed from request to review to implementation to closure

- Maintain agreed criteria for categorizing and classifying changes as well as for identifying change work products, testing levels, approval levels and tasks based on category and classification

- Build, manage and maintain the ongoing list of standard changes

- Manage, maintain and lead Change Advisory Board meetings

- Manage and maintain Forward Schedule of Changes

Service Initiation:

- Submitted RFCs (Request for Changes)

Service Delivery Channels:

- Completed RFCs

- Post implementation reviews

- Change notifications for approval/rejection

- Published forward schedule of changes

- Consulting support

Configuration and Asset Management

Description:

This service provides identification, tracking and control information about service configurations and assets in an accurate manner and makes that information available to all infrastructure services.

Service Scope:

All IT services in production, development or proposed.

Service Functions and Features:

- Plan, design, build, test, implement and maintain configuration and asset control management systems and solutions

- Identify and record configuration items used to support services

- Identify and record relationships between configuration items to support planning and impact analysis activities

- Provide information about configurations and assets to others through a formalized request mechanism

- Conduct configuration and asset audit activities to ensure reliability of information

- Plan, design, build, test, implement and maintain automated discovery technologies and automated linkages to populate configuration and asset control management systems and repositories

- Build and maintain blueprints and models of all services showing their components and relationships

- Maintain and manage information about warranties for physical assets

- Coordinate activities to dispose of assets upon formalized request

- Build, manage and maintain configuration and asset repository logical data schemas and designs

- Maintain accurate information about attributes of configurations and assets such as status, location, serial number, descriptions, owners, versions and relationships to other configuration items

- Provide status accounting to validate usage of configurations and assets

- Control updates to configuration and asset information through a formalized procedure that is linked with Change Control Services

- Build and maintain standards for naming and labeling configuration and asset components

- Provide and maintain configuration baselines

- Provide requirements and strategies to backup and recover configuration and asset management systems and data in the event of a major business disruption

Service Initiation:

- Approved work requests

- Escalated requests from the Service Desk

- Approved operational run procedures

Service Delivery Channels:

- Completed work requests

- Configuration and asset reports

- Configuration and asset repositories

- Consulting support

Lease and License Management

Description:

This service provides identification, reporting, tracking, and management of equipment leases and software licenses used to support IT and IT supported business services.

Service Scope:

All IT services in production or development.

Service Functions and Features:

- Plan, design, build, test, implement and maintain equipment leases, government issued licenses (i.e. FCC licenses) and software license management systems, data and solutions

- Identify and record equipment leases, government issued licenses and software licenses used to support services

- Identify required constraints on equipment leases and software licenses such as usage restrictions, start dates, end dates, and return restrictions

- Manage and maintain equipment lease, government licenses and software license cost information

- Provide information about cost implications, penalties and charges for changes to leases and licenses

- Track and report lease and license renewal and end-of-life dates and provide notifications of these to appropriate personnel on a timely basis

175

- Monitor and control usage of software licenses identifying thresholds and actions to be taken

- Proactively identify leases and licenses not used, no longer needed or would need to have their usage levels decreased or increased

- Monitor and control maintenance activities for leased equipment ensuring that equipment is maintained within lease requirements

Service Initiation:

- Approved work requests

- Escalated requests from the Service Desk

- Approved operational run procedures

Service Delivery Channels:

- Completed work requests

- Lease usage reports

- Lease compliance status reports

- Consulting support

Access and Security Management

Description:

This service supports the corporate security policy by providing controls over access to service infrastructure components and protecting services from unauthorized access or use.

Service Scope:

All IT services in production or development and the physical locations where those services are delivered and supported.

Service Functions and Features:

- Provide monitoring and reporting over security events ensuring all recognized events are logged, escalated and reported on

- Investigate and report on security breaches

- Provide Identity Management support creating, modifying and removing account IDs, passwords and account records maintenance activities

- Provide management and maintenance of certificates

- Manage and maintain synchronization of passwords for single sign on solutions

- Plan, design, build, test, implement and customize hardware, software and directories used to support security solutions and automate enforcement of corporate security policies

- Build, manage and maintain access control lists, grouping assignments, rights assignments and access profiles

- Provide and maintain protection over viruses and intrusion threats

- Build, manage and maintain physical security solutions such as badge creation and maintenance, camera surveillance, physical access management to secure areas and physical authentication such as thumb or retina biometrics access

- Provide protocol management services such as encryption support, VPN access and secured transmission of data and files

- Provide presence management services such as federated security management and trusted partner support

- Plan, design, build, implement, test and maintain internet firewalls, internet port access and proxies

- Plan, design, build, implement, test and maintain standards for application security, security models, application security API interfaces and security validation

- Conduct regular audits and tests for intrusions, vulnerability and security risk exposures

- Support internal and external corporate audit efforts as well as validate security compliance with regulatory and industry controls

- Provide security patch management services to notify and communicate needed patches, validate that patches have been successfully applied, identifies current versus desired patch levels and provides research into available security patches

- Plan, design, build, test, implement and maintain an integrated set of system directories

- Provide security consulting services to assist with security design over new and changing services

Service Initiation:

- Approved work requests

- Escalated incidents and requests from the Service Desk

- Approved operational run procedures

- Security incidents, alarms and alerts

Service Delivery Channels:

- Completed work requests

- Security audit reports

- Physical surveillance monitoring

- Security plans and strategies

- Consulting support

Service Audit and Reporting

Description:

This service provides reports and information that identify how well services are delivered and whether they are meeting agreed targets and goals.

Service Scope:

This service includes all IT services that are running in production.

Service Functions and Features:

- Build and maintain a catalog of standard service reports that describe how well services are being delivered and the overall quality of service delivery and support activities

- Plan, design, build, test, implement and maintain strategies, procedures and technologies to assemble result data, create and distribute reports

- Plan, design, build, test, implement and maintain strategies and technologies to provide management dashboards and reporting web sites

- Build and maintain report distribution lists

- Plan, build, manage, maintain and coordinate a master set of service metrics, calculations and assumptions

- Provide consulting services to identify and implement a best practice set of key performance indicators, critical success factors and service targets for services

- Periodically conduct service audits to identify how well services are being delivered and potential opportunities for improvement

- Periodically conduct reporting reviews with management to identify improvements over the quality and use of reports provided

Service Initiation:

- Approved work requests

- Escalated requests from the Service Desk

- Approved operational run procedures

Service Delivery Channels:

- Completed work requests

- Service audit and quality reports

- Service reporting web sites

- Consulting support

IT Workforce Management

Description:

This service provides management, controls over IT workforce staffing as well as recruiting activities needed to support services.

Service Scope:

This service covers all company staff and operational locations that carry out delivery and support activities for IT and IT supported business services.

Service Functions and Features:

- Assign, coordinate and maintain workforce levels considering needs for roles, responsibilities, work shifts, weekends, holidays and recommended staffing levels

- Build and maintain an inventory of support roles and job descriptions that are aligned with corporate personnel and human resources policies

- Provide workforce scheduling support services to fulfill workforce needs upon request

- Support corporate and IT recruiting activities by coordinating service support needs, scheduling interviews and reviewing candidates for selection

- Plan, build and maintain strategies and technologies to communicate open job listings and postings

- Provide oversight and control services to conduct workforce personnel reviews and salary/compensation rankings

- Provide oversight and control services to ensure staff skills are at adequate levels and develop individualized staff training and improvement plans

- Provide oversight and administration over secretarial and administrative personnel including work assignments, managers assigned to and sourcing from staffing agencies and temp firms

- Provide services to schedule and communicate meetings for onsite and offsite locations, teleconferences and web casts

Service Initiation:

- Approved work requests

- Formal requests for job openings

- Approved operational procedures

Service Delivery Channels:

- Completed work requests

- Assigned IT service support and delivery staff

- Workforce scheduling and status reporting

- Job listings and postings

- Employee reviews

- Recruitment campaigns

- Interview results and summaries

- Consulting support

Procurement Support

Description:

This service provides support for timely procurement of hardware, software, network, consulting and service solutions from 3rd party suppliers that will be used to underpin IT and IT supported business services.

Service Scope:

This service includes all purchases on behalf of IT and IT supported business services in production, under development or proposed.

Service Functions and Features:

- Provide services to coordinate and support development of customer needs statements, Requests for Information (RFIs) and Requests for Proposals (RFPs) and Requests for Quotes (RFQs)

- Provide services to coordinate, review, select and approve vendor responses to RFIs, RFPs RFQs and service requests

- Provide services to coordinate activities with corporate procurement to purchase, create approved purchase orders and receive goods/services used to support services

- Proactively identify opportunities for lower pricing by consolidating procurement requests and identifying short lists of desired suppliers

- Coordinate procurements and delivery schedules with destination locations and installation teams to have the components delivered in a timely fashion and receive purchased goods and services

- Provide asset and configuration information on procured components to Configuration and Asset management services

Service Initiation:

- Approved procurement requests

- Requests escalated from the Service Desk

Service Delivery Channels:

- Completed procurement requests

- Delivered IT goods and services

- Formal procurement requests documents such as RFIs, RFPs, and RFQs

- Underpinning contracts (UCs)

- Consulting support

Process Management

Description:

This service provides oversight, controls and effective efficiency over processes used to support services.

Service Scope:

This includes all service support and service delivery processes that underpin IT production services, services in development and proposed services.

Service Functions and Features:

- Provide process ownership through design, implementation and continuous improvement of support processes

- Ensure all support processes are appropriately documented and communicated to service support staff

- Coordinate activities to measure and report on the value and effectiveness of processes used to support services

- Coordinate activities to take proactive actions in improving support processes

- Ensure appropriate ownership and accountability exists for each support process

- Coordinate activities to identify and approve key performance indicators and critical success factors for processes to be measured and reported on

- Conduct periodic audits to validate compliance with support processes

- Process exceptions and variations to established support processes

Service Initiation:

- Request for changes (RFCs) for process changes or new processes

- Approved work requests

- Incidents and requests escalated from the Service Desk

Service Delivery Channels:

- Process quality reports

- Process guides and policies

- Process information web sites

- Delivered IT goods and services

- Consulting support

Supplier Relationship Management

Description:

This service provides oversight and control over suppliers to ensure seamless quality delivery of services.

Service Scope:

This includes all suppliers that underpin IT services in production, services in development or services that are being proposed.

Service Functions and Features:

- Provide contract oversight and management services for each supplier supporting the services being delivered

- Develop and maintain a set of services and service delivery targets for each supplier and ensure that these are contracted for

- Provide regularly scheduled reporting over contractor service targets and whether they are being met

- Provide single point of contact liaison between the business and the supplier to escalate and resolve service issues and requirements

- Maintain and manage the supplier/business relationship

- Monitor supplier contracts for compliance with contracted services, service targets and overall corporate business policies

- Provide regularly scheduled communications with suppliers to review contracted service delivery quality, service issues and new service requirements

- Maintain and follow an agreed process for escalating and resolving contract disputes

- Provide oversight and coordination services to transfer contracted services from one provider to another

- Manage supplier invoices for proper charging and payment and resolve any invoice/billing discrepancies

Service Initiation:

- Approved work requests

- Incidents and requests escalated from the Service Desk

- Approved operational procedures

Service Delivery Channels:

- Supplier performance reports

- Completed Request for Changes (RFCs) for underpinning contract changes

- Supplier liaisons

- Consulting support

Chapter

9

Hosting and Cloud Support Services

This category includes IT services that bundle support for hosting and operating an IT infrastructure that underpins one or more IT or business services. Services may be provided via cloud and virtualization technology delivery channels.

- Basic Support Service
- Infrastructure As A Service (IaaS)
- Platform As A Service (PaaS)
- Application As A Service (AaaS)
- Software as a Service (SaaS)
- Network as a Service (NaaS)
- Security as a Service
- Storage as a Service
- Equipment as a Service
- Secure Controlled Infrastructure Facility (SCIF)

Services can underpin any IT or business service running in production or under development. They may include many of the IT support services as features. Bundling of services might provide options for increasingly wider levels of scope as follows:

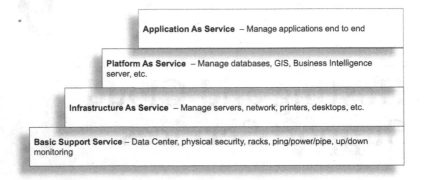

Figure 5: Bundling and Tiering Of Cloud-based Hosting Services

IT needs to be aware of a key requirement for customers when providing options such as those described above. Customers should not be able to pick and choose a hosting option by itself when there are dependencies on other hosting options they have not selected!

A prime example is Application as a Service. Customers should not be allowed to only choose that option without also selecting the IaaS and Basic Support Service options. The reason is simple. IT cannot be held responsible for service quality of the applications if they do not own responsibility for managing the underlying hardware and operating software that those applications run on.

One last consideration for these kinds of services is how they are bundled and priced for customers. A typical IT approach is to meter for every piece and part of the service (e.g. charge for processing, storage, network by amounts used). This approach can quickly get frustrating for both users and the IT organization as there can be a lot of overhead to meter, report and charge on usage.

An alternative is to bundle many complex items being provided into capability packages. With this approach, customers are presented with "blocks" of service features and functions for a single price. As an example for providing infrastructure services, you might offer the following:

Table 2: Example Offerings For Cloud-based Hosting Services

Offering	Description	Pricing
Basic	Single server with 128GB memory, 100GB storage, high speed network connection, basic support	$30.00 per month
Silver	Dual processor with 256GB memory, 500GB storage, high speed network connection, OS support included	$90.00 per month
Gold	Quad processor with 512GB Memory, 1TB of storage, high speed network connection, full operational support provided	$150.00 per month
Storage Plus	Additional blocks of 128GB storage	$20.00 per month for each block added
Recovery Plus	Basic, Silver or Gold provided with fully redundant infrastructure recoverable within minutes	Additional $50.00 per month

The above approach is much easier to manage and provides a more clear transparency in pricing to customers.

Basic Support Service

Description:

This provides foundational bare-bones data center facilities, service desk operations and a network connection. Customers are responsible for managing everything else themselves. Some considerations need to be taken into account such as how to handle on-site maintenance and repair for customer owned equipment or data center access to customer repair vendors.

Service Scope:

Services are provided for all company employees and include cradle-to-grave support from initial desktop procurement to installation, setup, ongoing support and eventual removal.

Service Functions and Features:

- Provide request fulfillment processing for client requests for desktops

- Provide desktop implementation, schedule coordination, move, add, change and disposal services when requested

- Obtain and validate requests for desktop security IDs, passwords and access profiles

- Provide access to Service Desk facilities to resolve desktop incidents or provide how-to support

- Provide planning, consulting and coordination services to establish, move or remove office LAN, printer and desktop facilities

- Provide planning, consulting and coordination services to upgrade desktop hardware and software configurations to newer versions

- Manage and maintain pick list of desktop assets slated for renewal or retirement and coordinate activities with clients to initiate those actions

- Provide asset and cost information to client managers and supervisors when requested

- Provide and maintain standard catalog of approved desktop hardware and software configurations available for use

- Identify client required pre-requisites for obtaining desktop support services as well as any post requisite activities once services are delivered

Service Initiation:

- Approved work requests

- Incidents and requests escalated from the Service Desk

- Approved operational procedures

Service Delivery Channels:

- Completed work requests

- Service Desk support

- Consulting support

Infrastructure As A Service (IAAS)

Description:

This adds on the management and maintenance of servers, printers, network or desktops (virtualized or non-virtualized) and their core operating systems. Customers are still responsible for their own add-on software and applications. Some considerations for how to configure OS software to accommodate customer software and applications needs to be taken into account.

Service Scope:

Services are provided for all company employees and include cradle-to-grave support from initial desktop procurement to installation, setup, ongoing support and eventual removal.

Service Functions and Features:

- Provide request fulfillment processing for customer hosting requests

- Provide due diligence services to assess client hosting requirements and estimate hosting costs and charges

- Provide and maintain hardware and software configurations required to meet hosting business needs

- Provide network connectivity and interfaces to allow clients to access the hosted infrastructure

- Provide security mechanisms, access IDs, passwords and profiles to allow clients to access the hosted infrastructure as well as to protect the hosting environment itself

- Provide a billing dashboard, metering, or other means for customers to self-manage and control their costs and resource usage.

- Plan, design, build, test, implement and maintain the overall hosting architecture to accommodate multiple hosting operations and services from the hosting services infrastructure

- Provide access to Service Desk facilities to resolve incidents or provide how-to support for hosting services

- Provide planning, consulting and coordination services to upgrade hosting services, facilities and arrangements when requested

- Provide and maintain standard catalog of approved hosting solutions and configurations available for use

- Identify client required pre-requisites for obtaining hosting services as well as any post requisite activities once services are delivered

- Maintain capacity and performance capabilities to meet agreed hosting business needs

- Provide availability management and service continuity activities to protect hosting services in the event of a major business disruption

- Provide standard set of reports on hosting services delivery activities, incidents, issues and status on a scheduled basis

Service Initiation:

- Negotiated hosting agreement or contract

- Approved operational procedures

Service Delivery Channels:

- Internet

- Intranet

- Cloud infrastructure

- Virtualized infrastructure

- Non-virtualized (physical) infrastructure

- Onsite housing

- Co-location housing

- Remote hosting location

- Consulting support

Platform As A Service (PAAS)

Description:

This adds on the management and maintenance of specialized platforms (e.g. database systems, GIS systems, CRM systems). Scope includes all hardware, software and networking that makes up the platform. The customer is still responsible for any applications that make calls to or run on the platforms being supported.

Service Scope:

Services are provided for all company employees and include cradle-to-grave support from initial desktop procurement to installation, setup, ongoing support and eventual removal.

Service Functions and Features:

- Provide request fulfillment processing for customer hosting requests

- Provide due diligence services to assess client hosting requirements and estimate hosting costs and charges

- Provide and maintain hardware and software configurations required to meet hosting business needs

- Provide network connectivity and interfaces to allow clients to access the hosted infrastructure

- Provide security mechanisms, access IDs, passwords and profiles to allow clients to access the hosted infrastructure as well as to protect the hosting environment itself

- Provide a billing dashboard, metering, or other means for customers to self-manage and control their costs and resource usage.

- Plan, design, build, test, implement and maintain the overall hosting architecture to accommodate multiple hosting operations and services from the hosting services infrastructure

- Provide access to Service Desk facilities to resolve incidents or provide how-to support for hosting services

- Provide planning, consulting and coordination services to upgrade hosting services, facilities and arrangements when requested

- Provide and maintain standard catalog of approved hosting solutions and configurations available for use

- Identify client required pre-requisites for obtaining hosting services as well as any post requisite activities once services are delivered

- Maintain capacity and performance capabilities to meet agreed hosting business needs

- Provide availability management and service continuity activities to protect hosting services in the event of a major business disruption

- Provide standard set of reports on hosting services delivery activities, incidents, issues and status on a scheduled basis

Service Initiation:

- Negotiated hosting agreement or contract

- Approved operational procedures

Service Delivery Channels:

- Internet

- Intranet

- Cloud infrastructure

- Virtualized infrastructure

- Non-virtualized (physical) infrastructure

- Onsite housing

- Co-location housing

- Remote hosting location

- Consulting support

Application As A Service (AAAS)

Description:

This adds on the management and maintenance of customer applications all the way from the hardware they run on through maintenance and modification of application code and systems. Customers that go with this option are expecting that IT essentially is running everything and that they only use the systems that are being managed.

Service Scope:

Services are provided for all company employees and include cradle-to-grave support from initial desktop procurement to installation, setup, ongoing support and eventual removal.

Service Functions and Features:

- Provide request fulfillment processing for customer hosting requests

- Provide due diligence services to assess client hosting requirements and estimate hosting costs and charges

- Provide and maintain hardware and software configurations required to meet hosting business needs

- Provide network connectivity and interfaces to allow clients to access the hosted infrastructure

- Provide security mechanisms, access IDs, passwords and profiles to allow clients to access the hosted infrastructure as well as to protect the hosting environment itself

- Provide a billing dashboard, metering, or other means for customers to self-manage and control their costs and resource usage.

- Plan, design, build, test, implement and maintain the overall hosting architecture to accommodate multiple hosting operations and services from the hosting services infrastructure

- Provide access to Service Desk facilities to resolve incidents or provide how-to support for hosting services

- Provide planning, consulting and coordination services to upgrade hosting services, facilities and arrangements when requested

- Provide and maintain standard catalog of approved hosting solutions and configurations available for use

- Identify client required pre-requisites for obtaining hosting services as well as any post requisite activities once services are delivered

- Maintain capacity and performance capabilities to meet agreed hosting business needs

- Provide availability management and service continuity activities to protect hosting services in the event of a major business disruption

- Provide standard set of reports on hosting services delivery activities, incidents, issues and status on a scheduled basis

Service Initiation:

- Negotiated hosting agreement or contract

- Approved operational procedures

Service Delivery Channels:

- Internet

- Intranet

- Cloud infrastructure

- Virtualized infrastructure

- Non-virtualized (physical) infrastructure

- Onsite housing

- Co-location housing

- Remote hosting location

- Consulting support

Software As A Service (SAAS)

Description:

This service providing specific software capabilities or products to customers.

Service Scope:

Services are provided for all company employees and include cradle-to-grave support from initial desktop procurement to installation, setup, ongoing support and eventual removal.

Service Functions and Features:

- Provide request fulfillment processing for customer hosting requests

- Provide due diligence services to assess client hosting requirements and estimate hosting costs and charges

- Provide and maintain hardware and software configurations required to meet hosting business needs

- Provide network connectivity and interfaces to allow clients to access the hosted infrastructure

- Provide security mechanisms, access IDs, passwords and profiles to allow clients to access the hosted infrastructure as well as to protect the hosting environment itself

- Provide a billing dashboard, metering, or other means for customers to self-manage and control their costs and resource usage.

- Plan, design, build, test, implement and maintain the overall hosting architecture to accommodate multiple hosting operations and services from the hosting services infrastructure

- Provide access to Service Desk facilities to resolve incidents or provide how-to support for hosting services

- Provide planning, consulting and coordination services to upgrade hosting services, facilities and arrangements when requested

- Provide and maintain standard catalog of approved hosting solutions and configurations available for use

- Identify client required pre-requisites for obtaining hosting services as well as any post requisite activities once services are delivered

- Maintain capacity and performance capabilities to meet agreed hosting business needs

- Provide availability management and service continuity activities to protect hosting services in the event of a major business disruption

- Provide standard set of reports on hosting services delivery activities, incidents, issues and status on a scheduled basis

Service Initiation:

- Negotiated hosting agreement or contract

- Approved operational procedures

Service Delivery Channels:

- Internet

- Intranet

- Cloud infrastructure

- Virtualized infrastructure

- Non-virtualized (physical) infrastructure

- Onsite housing

- Co-location housing

- Remote hosting location

- Consulting support

Network As A Service (NAAS)

Description:

This service providing network services such as WAN or LAN only for customers.

Service Scope:

Services are provided for all company employees and include cradle-to-grave support from initial desktop procurement to installation, setup, ongoing support and eventual removal.

Service Functions and Features:

- Provide request fulfillment processing for customer hosting requests

- Provide due diligence services to assess client hosting requirements and estimate hosting costs and charges

- Provide and maintain hardware and software configurations required to meet hosting business needs

- Provide network connectivity and interfaces to allow clients to access the hosted infrastructure

- Provide security mechanisms, access IDs, passwords and profiles to allow clients to access the hosted infrastructure as well as to protect the hosting environment itself

- Provide a billing dashboard, metering, or other means for customers to self-manage and control their costs and resource usage.

- Plan, design, build, test, implement and maintain the overall hosting architecture to accommodate multiple hosting operations and services from the hosting services infrastructure

- Provide access to Service Desk facilities to resolve incidents or provide how-to support for hosting services

- Provide planning, consulting and coordination services to upgrade hosting services, facilities and arrangements when requested

- Provide and maintain standard catalog of approved hosting solutions and configurations available for use

- Identify client required pre-requisites for obtaining hosting services as well as any post requisite activities once services are delivered

- Maintain capacity and performance capabilities to meet agreed hosting business needs

- Provide availability management and service continuity activities to protect hosting services in the event of a major business disruption

- Provide standard set of reports on hosting services delivery activities, incidents, issues and status on a scheduled basis

Service Initiation:

- Negotiated hosting agreement or contract

- Approved operational procedures

Service Delivery Channels:

- Internet

- Intranet

- Cloud infrastructure

- Virtualized infrastructure

- Non-virtualized (physical) infrastructure

- Onsite housing

- Co-location housing

- Remote hosting location

- Consulting support

Security As A Service

Description:

This service providing network services such as WAN or LAN only for customers.

Service Scope:

Services are provided for all company employees and include cradle-to-grave support from initial desktop procurement to installation, setup, ongoing support and eventual removal.

Service Functions and Features:

- Provide request fulfillment processing for customer hosting requests

- Provide due diligence services to assess client hosting requirements and estimate hosting costs and charges

- Provide and maintain hardware and software configurations required to meet hosting business needs

- Provide network connectivity and interfaces to allow clients to access the hosted infrastructure

- Provide security mechanisms, access IDs, passwords and profiles to allow clients to access the hosted infrastructure as well as to protect the hosting environment itself

- Provide a billing dashboard, metering, or other means for customers to self-manage and control their costs and resource usage.

211

- Plan, design, build, test, implement and maintain the overall hosting architecture to accommodate multiple hosting operations and services from the hosting services infrastructure

- Provide access to Service Desk facilities to resolve incidents or provide how-to support for hosting services

- Provide planning, consulting and coordination services to upgrade hosting services, facilities and arrangements when requested

- Provide and maintain standard catalog of approved hosting solutions and configurations available for use

- Identify client required pre-requisites for obtaining hosting services as well as any post requisite activities once services are delivered

- Maintain capacity and performance capabilities to meet agreed hosting business needs

- Provide availability management and service continuity activities to protect hosting services in the event of a major business disruption

- Provide standard set of reports on hosting services delivery activities, incidents, issues and status on a scheduled basis

Service Initiation:

- Negotiated hosting agreement or contract

- Approved operational procedures

Service Delivery Channels:

- Internet

- Intranet

- Cloud infrastructure

- Virtualized infrastructure

- Non-virtualized (physical) infrastructure

- Onsite housing

- Co-location housing

- Remote hosting location

- Consulting support

Storage As A Service

Description:

This service providing customers only with storage capabilities.

Service Scope:

Services are provided for all company employees and include cradle-to-grave support from initial desktop procurement to installation, setup, ongoing support and eventual removal.

Service Functions and Features:

- Provide request fulfillment processing for client requests for desktops

- Provide desktop implementation, schedule coordination, move, add, change and disposal services when requested

- Obtain and validate requests for desktop security IDs, passwords and access profiles

- Provide access to Service Desk facilities to resolve desktop incidents or provide how-to support

- Provide planning, consulting and coordination services to establish, move or remove office LAN, printer and desktop facilities

- Provide planning, consulting and coordination services to upgrade desktop hardware and software configurations to newer versions

- Manage and maintain pick list of desktop assets slated for renewal or retirement and coordinate activities with clients to initiate those actions

- Provide asset and cost information to client managers and supervisors when requested

- Provide and maintain standard catalog of approved desktop hardware and software configurations available for use

- Identify client required pre-requisites for obtaining desktop support services as well as any post requisite activities once services are delivered

Service Initiation:

- Approved work requests

- Incidents and requests escalated from the Service Desk

- Approved operational procedures

Service Delivery Channels:

- Completed work requests

- Service Desk support

- Consulting support

Equipment As A Service

Description:

This service provides customers with the ability to procure and run on hardware that is owned by the IT department versus customer owned equipment. This provides advantages in better support and pricing (assuming IT has negotiated bulk buying discounts).

Service Scope:

Services are provided for all company employees and include cradle-to-grave support from initial desktop procurement to installation, setup, ongoing support and eventual removal.

Service Functions and Features:

- Provide request fulfillment processing for customer hosting requests

- Provide due diligence services to assess client hosting requirements and estimate hosting costs and charges

- Provide and maintain hardware and software configurations required to meet hosting business needs

- Provide network connectivity and interfaces to allow clients to access the hosted infrastructure

- Provide security mechanisms, access IDs, passwords and profiles to allow clients to access the hosted infrastructure as well as to protect the hosting environment itself

- Provide a billing dashboard, metering, or other means for customers to self-manage and control their costs and resource usage.

- Plan, design, build, test, implement and maintain the overall hosting architecture to accommodate multiple hosting operations and services from the hosting services infrastructure

- Provide access to Service Desk facilities to resolve incidents or provide how-to support for hosting services

- Provide planning, consulting and coordination services to upgrade hosting services, facilities and arrangements when requested

- Provide and maintain standard catalog of approved hosting solutions and configurations available for use

- Identify client required pre-requisites for obtaining hosting services as well as any post requisite activities once services are delivered

- Maintain capacity and performance capabilities to meet agreed hosting business needs

- Provide availability management and service continuity activities to protect hosting services in the event of a major business disruption

- Provide standard set of reports on hosting services delivery activities, incidents, issues and status on a scheduled basis

Service Initiation:

- Negotiated hosting agreement or contract

- Approved operational procedures

Service Delivery Channels:

- Internet

- Intranet

- Cloud infrastructure

- Virtualized infrastructure

- Non-virtualized (physical) infrastructure

- Onsite housing

- Co-location housing

- Remote hosting location

- Consulting support

Secure Controlled Infrastructure Facility (SCIF)

Description:

This service providing customers with the ability to utilize any of the above services in a physically secured and controlled operating facility.

Service Scope:

Services are provided for all company employees and include cradle-to-grave support from initial desktop procurement to installation, setup, ongoing support and eventual removal.

Service Functions and Features:

- Provide request fulfillment processing for customer hosting requests

- Provide due diligence services to assess client hosting requirements and estimate hosting costs and charges

- Provide and maintain hardware and software configurations required to meet hosting business needs

- Provide network connectivity and interfaces to allow clients to access the hosted infrastructure

- Provide security mechanisms, access IDs, passwords and profiles to allow clients to access the hosted infrastructure as well as to protect the hosting environment itself

219

- Provide a billing dashboard, metering, or other means for customers to self-manage and control their costs and resource usage.

- Plan, design, build, test, implement and maintain the overall hosting architecture to accommodate multiple hosting operations and services from the hosting services infrastructure

- Provide access to Service Desk facilities to resolve incidents or provide how-to support for hosting services

- Provide planning, consulting and coordination services to upgrade hosting services, facilities and arrangements when requested

- Provide and maintain standard catalog of approved hosting solutions and configurations available for use

- Identify client required pre-requisites for obtaining hosting services as well as any post requisite activities once services are delivered

- Maintain capacity and performance capabilities to meet agreed hosting business needs

- Provide availability management and service continuity activities to protect hosting services in the event of a major business disruption

- Provide standard set of reports on hosting services delivery activities, incidents, issues and status on a scheduled basis

Service Initiation:

- Negotiated hosting agreement or contract

- Approved operational procedures

Service Delivery Channels:

- Internet

- Intranet

- Cloud infrastructure

- Virtualized infrastructure

- Non-virtualized (physical) infrastructure

- Onsite housing

- Co-location housing

- Remote hosting location

- Consulting support

Chapter

10

IT Business Facing Services

This category includes IT services that support business functions, clients, business partners and customers.

- Desktop Support
- Data Warehousing and Business Intelligence
- Internet Telephony Service
- Email and Messaging
- Service Introduction
- Other IT Business Facing Services

Desktop Support

Description:

This service provides a single point of contact for all desktop, and associated peripheral needs of company employees.

Service Scope:

Services are provided for all company employees and include cradle-to-grave support from initial desktop procurement to installation, setup, ongoing support and eventual removal.

Service Functions and Features:

- Provide request fulfillment processing for client requests for desktops

- Provide desktop implementation, schedule coordination, move, add, change and disposal services when requested

- Obtain and validate requests for desktop security IDs, passwords and access profiles

- Provide access to Service Desk facilities to resolve desktop incidents or provide how-to support

- Provide planning, consulting and coordination services to establish, move or remove office LAN, printer and desktop facilities

- Provide planning, consulting and coordination services to upgrade desktop hardware and software configurations to newer versions

- Manage and maintain pick list of desktop assets slated for renewal or retirement and coordinate activities with clients to initiate those actions

- Provide asset and cost information to client managers and supervisors when requested

- Provide and maintain standard catalog of approved desktop hardware and software configurations available for use

- Identify client required pre-requisites for obtaining desktop support services as well as any post requisite activities once services are delivered

Service Initiation:

- Approved work requests

- Incidents and requests escalated from the Service Desk

- Approved operational procedures

Service Delivery Channels:

- Completed work requests

- Service Desk support

- Consulting support

Data Warehousing and Business Intelligence

Description:

This service provides delivery, storage and receipt of business information assets and makes these visible and available to applications and functions that support the business.

Service Scope:

Services are provided for all company employees and executive management.

Service Functions and Features:

- Plan, design, build, test, implement and maintain data warehouse strategies, tools, services and overall architecture

- Develop and maintain a set of data meta-views and schemas for storing and accessing data in the warehouse

- Provide facilities to capture, extract, cleanse, validate, replicate, manipulate, transform and load data for retrieval and access

- Provide facilities to add custom processing and stored logic routines to generate additional value data from that originally collected and captured

- Provide request services and access facilities to mine and retrieve data upon scheduled or ad-hoc request

- Provide consulting services to assist the business in query analysis and mining data for specialized information needs

- Provide backup and recovery services for the data warehouse to maintain its availability and prevent data loss in the event of a major business disruption

- Provide and maintain security mechanisms to prevent data from unauthorized access and loss of accuracy

- Provide and maintain integration links for automated collection or distribution of data between warehouse facilities and applications

- Provide and maintain capacity and cost information for building, operating and charging for data warehousing services

- Provide consulting services for tuning, requirements analysis, modeling, and prototyping services to assist with building information solutions

Service Initiation:

- Approved work requests

- Requests escalated from the Service Desk

- Ad-hoc report requests

- Approved operational procedures

- Standard business intelligence reports

Service Delivery Channels:

- Completed work requests

- Completed requests escalated from the Service Desk

- Delivered Ad-hoc reports

- Delivered standard business intelligence reports

- Consulting support

Internet Telephony Service

Description:

This service provides a full functioning telephony service delivered over the company internet or intranet instead of through traditional telephone services.

Service Scope:

Services are provided for all company employees.

Service Functions and Features:

- Provide a standard procurement mechanism for telephone devices, telephone numbers, telephone functions and features

- Provide available dial tone

- Provide and maintain high levels of voice quality

- Provide options of dialing plans for differing phone needs (e.g. a phone located in a public lobby versus an administrator phone)

- Provide standard and customized calling features such as Call Waiting, 3-way calling, Voice mail and Conferencing

- Process and coordinate requests for telephony service changes in a timely manner

- Provide service dispatch and repair services when telephony incidents are escalated from the Service Desk

- Provide billing services for telephony phone usage to recover infrastructure service costs

- Design, build, test, implement, deploy and maintain the infrastructure that underpins telephony services

- Design, build, test, implement, deploy and maintain the service continuity infrastructure that underpins telephony services in the event of a major business disruption

- Provide standard and ad-hoc reports that document usage of telephony services

- Decommission telephones and shutdown telephony service when requested for employees or business units

- Provide moves, adds and changes to telephony services based on requests

- Maintain overall employee satisfaction levels for telephony services provided and take proactive actions to improve services as needed

Service Initiation:

- Approved work requests

- Requests and incidents escalated from the Service Desk

- Approved operational procedures

Service Delivery Channels:

- Delivered telephony services via approved company telephone devices and systems

- Completed work requests

- Completed requests escalated from the Service Desk

- Resolved incidents escalated from the Service Desk

- Consulting support

Email and Messaging

Description:

This service provides delivery, storage and receipt of electronic and voice messages between company employees.

Service Scope:

Services are provided for all company employees and include messages delivered via Email, Instant Messaging, Text messaging and Voice Mail.

Service Functions and Features:

- Send and receive email, voice mail and instant messages, both internally and via the Internet

- Provide calendar feature with scheduling capability for appointments and reminders

- Provide an easily accessible global address list of email addresses and "white page" information

- Provide mailbox storage space in fixed sizes to hold and store messages per identified business requirements

- Send and receive file attachments (such as Word documents, Excel spreadsheets, etc.) with messages being sent or retrieved

- Provide the ability to create mail distribution groups holding a number of e-mail addresses for easy distribution

- Archive, purge and delete Email and calendar items per corporate legal and business requirements

- Provide access to messaging services over the internet via company approved browser solutions

- Provide shared messages, contacts, and calendar items through Public Folders

- Protect messages and services with anti-virus and anti-spam solutions

- Provide encryption capabilities to secure sensitive messages

Service Initiation:

- Approved work requests

- Requests and incidents escalated from the Service Desk

- Approved operational procedures

Service Delivery Channels:

- Delivered messages via desktop, laptop, voice, tablet and mobile devices

- Completed work requests

- Completed requests escalated from the Service Desk

- Resolved incidents escalated from the Service Desk

- Consulting support

Service Introduction

Description:

This service provides a single point of contact to business customers for provisioning, building and transitioning new or changed services into production.

Service Scope:

Any service solution targeted for production operations.

Service Functions and Features:

- Provide request fulfillment processing for obtaining support for new or changed services

- Provide cradle to grave oversight, coordination and support to build and transition services into production

- Identify and establish criteria for successful service implementation

- Obtain all necessary approvals and signoffs for services being implemented or changed

- Build and maintain integrated project plans that identify all service build and transition activities along with key milestone targets and timeframes

- Hold and conduct regularly scheduled status meetings to track service introduction activities, progress and issues

- Coordinate activities and strategies to implement organizational aspects of new and changed services such as training to upgrade support skills, operational roles, responsibilities, supplier contracts, relationships and ongoing support organization

- Coordinate activities to consolidate service requirements, needs and costs

- Coordinate activities to certify services for production operations

Service Initiation:

- Approved work requests

Service Delivery Channels:

- Completed work requests

- Consulting support

Other IT Business Services

Description:

These services represent IT applications and support of direct business functions such as corporate accounting, human resources, marketing, sales and other services unique to the industry that the business operates in.

A generic description template is presented below. This is followed by a list of typical business services by a number of industries:

Common Service Scope:

Any service solution targeted for production operations.

Common Service Functions and Features:

- Coordinate management and maintenance of applications that support all enterprise wide functions for the business service

- Satisfy requests for information and reports to support all enterprise wide functions for the business service

- Coordinate management and maintenance of applications to support key functions and features of the business function being addressed—these are unique for each business function and company— examples might look like the following for a service that supports the corporate legal department:

- Coordinate management and maintenance of applications that provide support for legal calendaring and litigation scheduling

- Coordinate management and maintenance of applications that provide support for handling and maintaining legal and official documents

- Manage and operate the legal department service desk

- Manage and operate the online law library

- Coordinate management and maintenance of applications that provide support for legal research activities

- Provide automated online services for management and maintenance of contract templates

- Provide automated online services for management and maintenance of versions of contract and legal documents

- Store, log and forward legal communications sent by outside parties through email or fax

Service Initiation:

- Approved work requests

Service Delivery Channels:

- Completed work requests

- Online services

- Consulting support

The following pages list general types of services by key industry areas that would fall under the category of IT Business Support Services:

General Corporate Services:

- General Ledger
- Accounts Receivable
- Accounts Payable
- Annual Report
- Financial Reporting
- Audit Support
- Payroll
- Human Resources
- Tax Accounting and Reporting

Manufacturing Services:

- Materials Management
- Receive Orders
- Fulfill Orders
- Shipping
- CAD Design Support
- Factory Floor Automation
- Manage Suppliers
- Supply Partner Relationship Management

Marketing Support Services:

- Market Research
- Market Planning
- Product History Maintenance
- Competitive Analysis
- Customer Insight
- Market Segmentation

Product Support Services:

- Contract Management
- Warranty Management
- Professional Services
- Product Education
- Repair History
- Repair Dispatching
- Repair Scheduling
- Parts and Supply Inventory Management

Sales Back Office Support Services:

- Order Management
- Billing
- Collections
- Commissions
- Customer Profile Management

Sales Front Office Support Services:

- Response Management
- Contact Management
- Lead Management
- Opportunity Management
- Sales Forecasting
- Channel Management
- Sale Management
- Pipeline Analysis

Customer Support Services:

- Service Analytics
- Solution Administration
- Case Management
- Self Service
- Customer Call Center
- Customer Survey Support Services

Product (or Service) Support Services:

- Contract Management
- Warranty Management
- Professional Services
- Product Education
- Repair History
- Repair Dispatching
- Repair Scheduling
- Parts and Supply Inventory Management
- Product Planning
- Product Description
- Development Lifecycle
- Campaign Planning

Procurement Support Services:

- Supplier Certification
- Contract Initiation
- Vendor Management
- Purchases
- Payments

Educational Services:

- Enrollment
- Registration
- Student Loan Processing
- Tuition Management
- Scholarship Management
- Course Scheduling
- Classroom Scheduling
- Pay Teaching Staff
- Course Catalog
- Research Support
- Library Management
- Student Housing Support
- Classroom Infrastructure Support
- Online Course Delivery and Management

Hospital Support Services:

- Admissions, Discharges and Transfers
- Lab Support
- Medical Records
- Pharmacy
- Radiology
- Emergency Room Scheduling
- Medical Staff Scheduling
- Patient Care Support
- Patient Billing
- Medical Supplies Management

Energy and Utilities Support Services:

- Customer Billing
- Order Fulfillment
- Meter Reading
- Infrastructure Repair and Management
- Power Generation
- Power Transmission
- Power Retailing
- Energy Development
- Environmental and Safety Support
- Load Profiling
- Fleet Maintenance
- Regulatory Control Support

Financial Trading Support Services:

- Portfolio Tracking and Management
- Technical Analysis
- Fundamental Analysis
- Trading Support
- Trading Media Management
- Risk Management
- Currency Management
- News Feeds and Distribution

Insurance Services:

- Agent Commissions and Pay
- Claims Payment
- Policy Underwriting
- Policy Sales Support
- Fraud Detection Support
- Property Risk Management
- Catastrophe Risk Management and Reporting
- Community Program Support

Patent and Trademark Support Services:

- Patent Search
- Patent Applications Management
- Maintenance Fee Management
- Patent Publishing
- Patent Document Maintenance

Banking Services:

- Automated Teller Support
- Online Banking
- Account Management
- Retail Banking Support
- Commercial Banking Support
- Mortgage Application Support
- Credit Card Operations
- Regulatory Management and Reporting
- Credit Risk Management
- Teller Support

Chapter

11

Governing Services

Service Governance Considerations

A Governance process should be established to oversee and manage services. This process has the following goals:

- Ensure the service is delivering its functions and outputs to the satisfaction of its customers

- Manage and escalate service quality issues and complaints to their resolution

- Receive and provide due diligence for requests to add new services or change existing services

- Ensure services are aligned with business goals and objectives

- Expect that services will change over time. This is a natural consequence of changes in the business to stay competitive.

The following sections in this chapter provide one possible approach for governing services.

Service Governance Roles

Roles represent collections of tasks and skills to handle the sub-functions of the governance process. A role is not the same as a job description. The relationship between roles and jobs may fall into one or more of the following:

Table 3: Relating Roles To Jobs

Role is . . .	Example
Full Time	John is dedicated to the role full time—it is all he does in his job.
Part Time	John only performs the role for 10 hours per week—he has other tasks he performs not related to the role during the rest of the time.
Split Among Many Jobs	John, his manager and two co-workers all handle the role on a part time basis.
Handled By Multiple Jobs	John, his manager and two co-workers all handle the role on a full time basis.

Although the role is not always equal to a job, a role description provides valuable input to a formalized job description.

The following key roles are recommended for handling governance activities for ITSM services:

ITSM Steering Group

This group consists of senior executives and representatives from the business plus the ITSM Service Manager. It has the final say on service direction and service strategy. It acts as an escalation point for final decisions and investments in new services or changes to existing services. Key responsibilities of this group are:

- Sponsoring and championing the Service Portfolio and Service Catalog across the business enterprise

- Reviewing and approving major Service Portfolio and Service Catalog changes

- Reviewing and approving escalated Service Portfolio and Catalog exceptions

- Reviewing and making timely decisions on Service Portfolio or Catalog escalated issues

- Reviewing and providing feedback on the effectiveness of the ITSM Service Teams and the Service Governance process

Service Team

This group consists primarily of Service Owners. It is strongly recommended that each service in the portfolio and catalog have a Service Owner to act as a single point of contact for all aspects of the service. It is also okay if a Service Owner owns more than one service. Key responsibilities of this group are:

- Developing and updating the Service Portfolio and Service Catalog descriptions

- Managing, executing and maintaining the Service Governance Process

- Ensuring the Service Portfolio and Catalog is aligned with IT and business initiatives

- Reviewing and processing requested changes to the Service Portfolio or Catalog

- Reviewing and processing requests for Service Portfolio or Catalog exceptions and appeals

- Communicating and championing the Service Portfolio and Catalog across the business enterprise

- Educating and involving IT developers in the development and evolution of the Service Portfolio and Catalog

- Reviewing proposed request for changes (RFCs) to identify and flag impacts that they may have to the Service Portfolio or Service Catalog

Business Liaisons

This group represents IT services to the business community and its customers. Ideally, they provide a single point of contact to selected business units they represent for all the services in the Service Catalog. They communicate what services are available and how they may best be used. They also provide a first point of escalation for service quality issues when they arise and work with the Service Teams to get them resolved. Key responsibilities of this group are:

- Understanding the Service Governance Process

- Locally championing and supporting the Service Catalog and its Governance process

- Reviewing Service Catalog changes as necessary and providing feedback to the Service Team

- Leveraging the Service Catalog to the greatest extent possible when recommending IT solutions and services

- Submitting RFCs for Service Catalog changes and working in a cooperative manner to resolve service exceptions to the catalog

- Escalating service quality concerns and issues as necessary

Service Governance Process

The following process can be used for handling the typical activities involved with governing services. The overall service governance process can be pictured as follows:

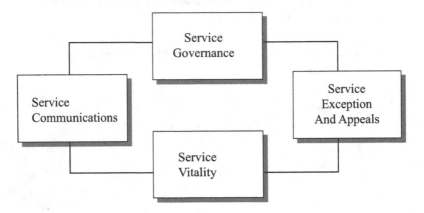

Figure 6: Service Governance Process

Service Governance

This provides a structured approach for reviewing and approving decisions for changing and evolving the Service Portfolio and the Service Catalog.

Service Exceptions and Appeals

This provides a means for escalating Service Portfolio and Catalog decisions for the use of non-conforming services to meet unique or changing IT and business requirements.

Service Vitality

This provides a way to incorporate new Service Portfolio and Service Catalog changes as a result of changing IT and business needs.

Service Communications

This provides a means for syndicating the Service Catalog as it evolves across the business enterprise.

The following pages show this process in more detail using LoveM style process flow charts.

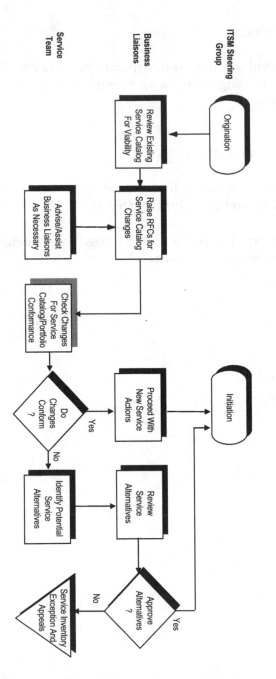

Figure 7: Service Governance Process

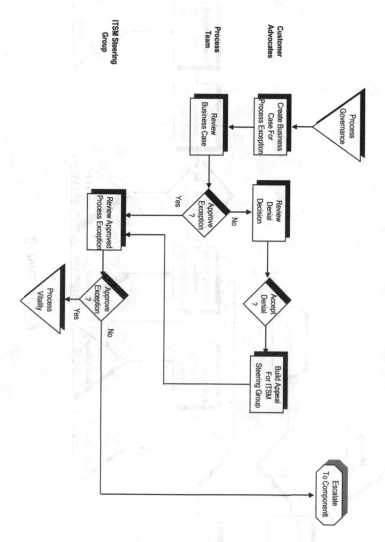

Process Exceptions and Appeals

Figure 8: Service Exception and Appeals Process

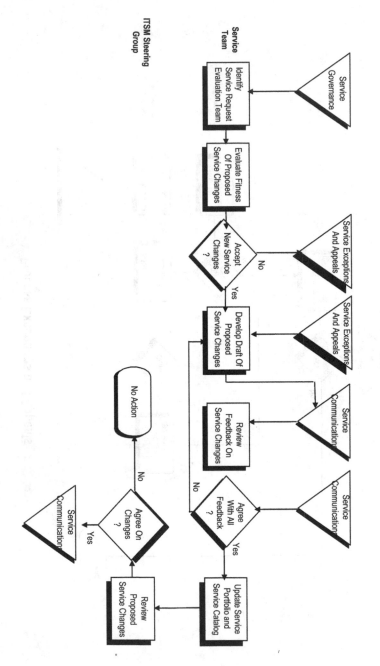

Figure 9: Service Vitality Process

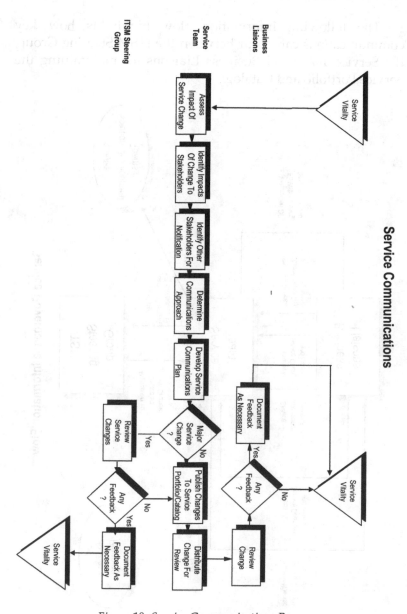

Figure 10: Service Communications Process

Service Governance Information Flow

The following information flow highlights how key communications can occur between the ITSM Steering Group, the Service Team and Business Liaisons for maintaining the Service Portfolio and Catalog:

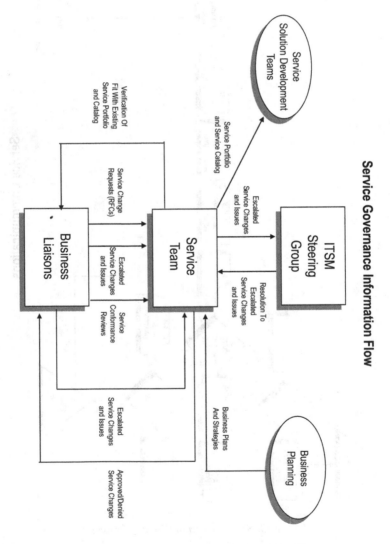

Figure 11: Service Governance Information Flow

Chapter

12

Service Implementation

General Implementation Considerations

This section describes some general considerations for implementing services. The current ITSM industry guidance is quite useful in this area as the guidance is now structured by a Service Lifecycle. The implementation effort for any given service can leverage this lifecycle in terms of organizing the needed tasks for implementing a service.

The key work phases for implementing a service, as presented here, follow the ITSM Service Lifecycle closely. The phases can be shown as follows:

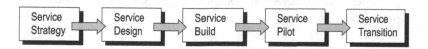

Figure 12: Lifecycle For Building A Service

The remaining two ITSM lifecycle phases: Service Operations and Continuous Service Improvement would occur in parallel once the service itself has completed these phases.

Staffing considerations for a service implementation effort will vary depending on the service being implemented. At a minimum, it is suggested that a Service Owner be assigned to the service along with a core team that will perform the implementation effort. An ITSM Service Manager should also be involved to make sure that the service is constructed in line with other IT services, processes and policies that may already be in place. Business liaisons should be involved to make sure the service is aligned with the needs of their customers as well.

Building the IT service should be initiated through the Service Governance Process described earlier in this book. Once approval for the new service (or changes to an existing service) has been obtained, the implementation activities described in this section can begin.

It is also recommended that each Work Phase (following the Service Strategy phase) operate with four project tracks. These are:

Process Track

Build and integrate the supporting processes that will underpin the service

Technology Track

Build and integrate the supporting technologies that will underpin the service

Organization Track

Build and integrate the supporting organization that will underpin the service as well as ensure training, communications and acceptance of the service

Governance Track

Build and integrate service quality measures, reporting and alignment with IT and business objectives for the service

Management Track

Perform Project Management duties and oversight to ensure the service is constructed in scope, on time and within budget

These tracks all operate simultaneously within each Work Phase of the implementation effort. A brief overview of each Work Phase is as follows:

Service Strategy

This phase covers tasks to establish the overall service scope, sourcing, business outcomes, costs and revenue. It also puts the service implementation team into place. At the end of this phase, approval is given to begin design work on the service (or changes if an existing service).

Service Design

This phase covers tasks to design the overall service functions and features. Design scope includes all processes, technologies, organization, data and reporting that will underpin the service. At the end of this phase, approval is given to begin investment in building the service.

Service Build

This phase covers tasks to build the overall service by creating the needed service assets (processes, technologies, people, etc.) that will underpin it. At the end of this phase, approval is given to begin investment in piloting the service with a selected set of customers, business units or end users.

Service Pilot

This phase covers tasks to execute the Service Pilot. At the end of this phase, approval is given to begin service transition activities to roll the service out to the rest of its intended customers.

Service Transition

This phase covers tasks to transition and roll out the service to the remaining customers, business units and end users. At the end of this phase, final approval is given to begin production operations for the service.

A detailed service implementation plan is provided in the next section of this chapter. It is intended to be a generic service implementation plan that should work across the board for almost any service being implemented.

In actual use, you can take this plan as a starting point and fill it in with the specifics of the service you plan to implement. The Key Output column will provide the starting point for a detailed work breakdown structure that would go with your plan.

Detailed Service Implementation Plan

The following pages list a detailed generic implementation plan for implementing a service. This can be used as a starting point for your own implementation efforts. It should be customized for your specific needs based on the service being implemented.

Service Strategy Work Tasks

SERVICE STRATEGY	
Work Task	Key Output
Build Service Mission and Charter	Service Charter
Identify Key Service Business Outcomes	Inventory Of Service Business Outcomes
Assign Service Development Team	Assigned Service Development Team
Develop Preliminary Scope Definition For Service	Preliminary Service Scope Definition
Identify Sourcing Strategy For Service	Service Sourcing Strategy
Identify Locations Where Service Will be Supported/Delivered	
Build Service Quality Plan	Service Quality Plan
Build Service Risk Plan	Service Risk Plan
Identify Key Service Integration Points	Inventory Of Key Service Integration Points
Develop Service Communication Plan	Service Communication Plan
Identify Needed Availability and Continuity Requirements	
Identify Needed Security Requirements	

SERVICE STRATEGY	
Work Task	Key Output
Estimate Service Demand	
Estimate One-Time Costs For Building Service	One-Time Cost Estimates
Estimate Ongoing Costs For Maintaining Service	Ongoing Cost Estimates
Estimate Revenue benefits From Service	Service Revenue Estimates
Build Service Implementation Plan	
Obtain Approval To Develop Service	Management Approval
Build Service Mission and Charter	Service Charter
Identify Key Service Business Outcomes	Inventory Of Service Business Outcomes
Assign Service Development Team	Assigned Service Development Team

Service Design Work Tasks

SERVICE DESIGN	
Work Task	Key Output
Process Track	
Identify Current Service Process Assets That Can Be Leveraged	
Integrate Service Process Assets	Service Support and Delivery Process Models
Design Service Workflows and Procedures	Service Designs and Blueprints
Identify Current Service Process Assets That Can Be Leveraged	Inventory of Gaps To Overcome For Service
Integrate Service Process Assets	
Technology Track	

SERVICE DESIGN	
Work Task	Key Output
Identify Service Support and Delivery Technology Requirements	Inventory of Service Technology Requirements
Design Service Input Forms/Screens	Service Input Forms and Screen Templates
Design Service Output Reports/ Screens	Service Output Forms and Screen Templates
Design Needed Availability and Service Continuity Functions	Integrated Service Continuity Plan
Design Needed Service Security Functions	Service Security Designs
Identify Current Technology Assets That Can Be Leveraged	
Develop Technology Gap Analysis	Technology Gap Analysis
Select Vendor Tools If Needed	Selected Vendor Tools
Develop Service Technology Support Strategy	Service Support Plan
Estimate Capacity Of Service Technology Assets Needed	Service Capacity Plan
Integrate Service Technology Assets	Service Architecture
Build and Develop Service Configuration and Asset Model	Service Configuration Model
Develop Service Technology Implementation Plan	Technology Implementation Plan
Organization Track	
Develop Service Execution Organization Strategy	
Develop Roles To Support/Deliver Service	Service Roles
Identify Needed Skills and Skill Levels For Each Role	Skills Descriptions

SERVICE DESIGN	
Work Task	Key Output
Develop Role Responsibilities For Service	Service Responsibilities
Develop Future State Service Support/Delivery Organization	Service Organization Model
Conduct Organization Gap Analysis	Organization Gap Analysis
Update Job Descriptions With New Role Information As Needed	Updated Job Descriptions
Estimate Capacity Of Service Workforce Assets Needed	Estimated Staffing Levels
Build Organization Transition Plan	Organization Transition Plan
Governance Track	
Develop Service Feature and Functional Specifications	
Determine Service Targets and Metrics	Service Targets and Metrics
Create Service Description	Detailed Service Description
Update Service Portfolio With Service Description	Updated Service Portfolio
Update Service Catalog With Service Description	Updated Service Catalog
Develop Service Policies	Service Policies
Develop Service Release Strategy	Service Release Plan
Refine Service Risk Plan	Service Risk Plan
Develop Service Quality Reports and Scorecards	Service Quality Reports
Develop Service Use Case Examples For Testing	Testing Use Cases
Establish Needed SLA and OLA Agreements	SLA and OLA Agreements

SERVICE DESIGN	
Work Task	Key Output
Establish Needed Supplier Contracts	Underpinning Contracts
Develop Service Cost Model	Service Cost Model
Complete Service Use Case Tests	Use Case Testing Results
Management Track	
Update Service Implementation Plan	Updated Project Plans
Manage Service Implementation Issues	Issues Inventory
Communicate Implementation Status	Implementation Status Reports

Service Build Work Tasks

SERVICE BUILD	
Work Task	Key Output
Process Track	
Develop Service Work Instructions	Service Procedures and Instructions
Draft Service Support/Delivery Operating Guide	Service Operating Guide
Technology Track	
Procure And Acquire Supporting Tools	Acquired Tools
Implement Supporting Tools And Changes	Installed Tools
Build Service Input Forms/Screens	
Build Service Output Reports/ Screens	
Customize Service Supporting Technologies As Needed	Customized Tools

SERVICE BUILD	
Work Task	Key Output
Update Service Configuration and Asset Models	Updated Service Configuration Models
Test Service Technology Architecture	Technology Test Results
Organization Track	
Develop Service Staffing Plan	Support Staff Plan
Assign Service Pilot Resources	Assigned Staffing Resources
Build Service Pilot Contact List	Pilot Contact List
Build Service Training Materials	Training Materials
Conduct Service Training For Pilot	Training Pilot
Governance Track	
Develop Service Transition Plans and Strategies	Service Transition Plan
Integrate Service Support Activities With Change Management	
Develop Service Release Packages	Service Release Packages
Select Service Pilot	Selected Service Pilot
Develop Service Pilot Validation Criteria	Pilot Validation Criteria
Develop Service Pilot Plan And Schedule	Pilot Plan and Schedule
Complete First Pass Service Validation Tests	Service Validation Unit Test Results
Package Service Documentation For Knowledge Management	Service Documentation Package
Management Track	
Update Service Implementation Plan	Updated Project Plans

SERVICE BUILD	
Work Task	Key Output
Manage Service Implementation Issues	Issues Inventory
Communicate Implementation Status	Implementation Status Reports

Service Pilot Work Tasks

SERVICE PILOT	
Work Task	Key Output
Process Track	
Execute Service Pilot	
Update Service Documentation As Needed	Updated Service Documentation
Technology Track	
Provide Tool Support For Service Pilot	
Maintain Supporting Tools For Service Pilot	Tooling Patches and Updates
Organization Track	
Monitor Pilot Roles And Responsibilities	
Update Service Organization Plan As Needed	Updated Service Organization Plan
Governance Track	
Produce Reporting For Service Pilot Quality	Pilot Quality Reports
Validate Pilot Results	Validated Pilot
Update Service Transition Plans	Updated Service Transition Plans
Management Track	
Update Service Implementation Plan	Updated Project Plans

SERVICE PILOT	
Work Task	Key Output
Manage Service Implementation Issues	Issues Inventory
Communicate Implementation Status	Implementation Status Reports

Service Transition Work Tasks

SERVICE DESIGN	
Work Task	Key Output
Process Track	
Support Service Procedures Pre and Post Transition	
Technology Track	
Support Service Technologies Pre and Post Transition	
Organization Track	
Execute Service Communication Strategies	
Support And Train Service Execution Organization	Trained Service Support Staff
Governance Track	
Negotiate and Agree SLA and OLA Agreements	Agreed SLAs and OLAs
Execute Service Transition Plans	
Monitor Service Transition Progress	
Validate Service Reporting In Place	
Report On Service Transition Progress	Transition Progress Status Reporting
Obtain Final Management Approval To Begin Production Operations	Management Approval

SERVICE DESIGN	
Work Task	Key Output
Management Track	
Manage Service Implementation Issues	Issues Inventory
Communicate Implementation Status	Implementation Status Reports

About the Author

Randy A. Steinberg has extensive IT Service Management and operations experience gained from many clients around the world. He authored the ITIL 2011 Service Operation book published worldwide. Passionate about game changing management practices within the IT industry, Randy is a hands-on IT Service Management expert helping IT organizations transform their IT infrastructure management strategies and operational practices to meet today's IT challenges.

Randy has served in IT leadership roles across many large government, health, financial, manufacturing and consulting firms including a role as Global Head of IT Service Management for a worldwide media company with 176 operating centers around the globe. He implemented solutions for one company that went on to win a Malcolm Baldrige award for their IT service quality. He continually shares his expertise across the global IT community frequently speaking and consulting with many IT technology and business organizations to improve their service delivery and operations management practices.

Randy can be reached at RandyASteinberg@gmail.com.